Transitioning to
Sustainable Life below Water

Transitioning to Sustainability Series: Volume 14

Series Editor: Manfred Max Bergman

Volumes in the series:

Volume 1: Transitioning to No Poverty
ISBN 978-3-03897-860-2 (Hbk);
ISBN 978-3-03897-861-9 (PDF)

Volume 2: Transitioning to Zero Hunger
ISBN 978-3-03897-862-6 (Hbk);
ISBN 978-3-03897-863-3 (PDF)

Volume 3: Transitioning to Good Health
and Well-Being
ISBN 978-3-03897-864-0 (Hbk);
ISBN 978-3-03897-865-7 (PDF)

Volume 4: Transitioning to Quality
Education
ISBN 978-3-03897-892-3 (Hbk);
ISBN 978-3-03897-893-0 (PDF)

Volume 5: Transitioning to Gender Equality
ISBN 978-3-03897-866-4 (Hbk);
ISBN 978-3-03897-867-1 (PDF)

Volume 6: Transitioning to Clean Water
and Sanitation
ISBN 978-3-03897-774-2 (Hbk);
ISBN 978-3-03897-775-9 (PDF)

Volume 7: Transitioning to Affordable and
Clean Energy
ISBN 978-3-03897-776-6 (Hbk);
ISBN 978-3-03897-777-3 (PDF)

Volume 8: Transitioning to Decent Work
and Economic Growth
ISBN 978-3-03897-778-0 (Hbk);
ISBN 978-3-03897-779-7 (PDF)

Volume 9: Transitioning to Sustainable
Industry, Innovation and Infrastructure
ISBN 978-3-03897-868-8 (Hbk);
ISBN 978-3-03897-869-5 (PDF)

Volume 10: Transitioning to Reduced
Inequalities
ISBN 978-3-03921-160-9 (Hbk);
ISBN 978-3-03921-161-6 (PDF)

Volume 11: Transitioning to Sustainable
Cities and Communities
ISBN 978-3-03897-870-1 (Hbk);
ISBN 978-3-03897-871-8 (PDF)

Volume 12: Transitioning to Responsible
Consumption and Production
ISBN 978-3-03897-872-5 (Hbk);
ISBN 978-3-03897-873-2 (PDF)

Volume 13: Transitioning to Climate Action
ISBN 978-3-03897-874-9 (Hbk);
ISBN 978-3-03897-875-6 (PDF)

Volume 14: Transitioning to Sustainable
Life below Water
ISBN 978-3-03897-876-3 (Hbk);
ISBN 978-3-03897-877-0 (PDF)

Volume 15: Transitioning to Sustainable
Life on Land
ISBN 978-3-03897-878-7 (Hbk);
ISBN 978-3-03897-879-4 (PDF)

Volume 16: Transitioning to Peace, Justice
and Strong Institutions
ISBN 978-3-03897-880-0 (Hbk);
ISBN 978-3-03897-881-7 (PDF)

Volume 17: Transitioning to Strong
Partnerships for the Sustainable
Development Goals
ISBN 978-3-03897-882-4 (Hbk);
ISBN 978-3-03897-883-1 (PDF)

Werner Ekau and Anna-Katharina Hornidge (Eds.)

Transitioning to
Sustainable Life below Water

Transitioning to Sustainability Series

MDPI • Basel • Beijing • Wuhan • Barcelona • Belgrade • Manchester • Tianjin • Tokyo • Cluj

EDITORS
Werner Ekau
Leibniz Centre for Tropical Marine
Research,
Bremen, Germany

Anna-Katharina Hornidge
German Development Institute
Bonn, Germany

EDITORIAL OFFICE
MDPI
St. Alban-Anlage 66
4052 Basel, Switzerland

For citation purposes, cite each article independently as indicated below:

Author 1, and Author 2. 2022. Chapter Title. In *Transitioning to Sustainable Life below Water.* Edited by Werner Ekau and Anna-Katharina Hornidge. Transitioning to Sustainability Series 14; Basel: MDPI, Page Range.

ISBN 978-3-03897-876-3 (Hbk)
ISBN 978-3-03897-877-0 (PDF)
ISSN: 2624-9324 (Print)
ISSN: 2624-9332 (Online)
doi:10.3390/books978-3-03897-877-0

Contents

About the Editors

Dr. Werner Ekau is Director of the Operational Centre Germany of the International Ocean Institute (IOI). From 2004 to 2017 he was a member and chair of the Committee of Directors. He is a fisheries biologist trained at Bochum, Hamburg and Kiel University and has worked in Antarctica, Argentina, and the Azores. Since 1991 his work has revolved around fish biology and ecology in tropical ecosystems and throughout the 1990s he was a coordinator for the bilateral co-operation in Marine Research between Brazil and Germany. His research interests lie in the wider field of "Adaptations of different life stages of fish to their environment and the consequences of climate change". This includes activities in growth and population dynamics; migration pattern of fish in coastal environments; ichthyoplankton; fisheries biology; ecology; and ecosystem responses to climate change. An important topic is his work is the ecosystem approach to fisheries management, by which he engaged for many years in the Large Marine Ecosystem community as well as the stimulation and enforcement of international cooperation and networking in marine science and governance. He is a member of the Group of Experts on Capacity Development of the Intergovernmental Oceanographic Commission of UNESCO.

Prof. Dr. Anna-Katharina Hornidge is Director of the German Development Institute (Deutsches Institut für Entwicklungspolitik or DIE) and Professor for Global Sustainable Development at the University of Bonn. In her research, Ms. Hornidge works on knowledges & innovation development for development, as well as questions of natural resources governance in agriculture and fisheries in Asia and Africa. Ms. Hornidge serves as expert advisor at national, EU and UN level: as Member of the German Advisory Council on Global Change of the German Government (WBGU), Co-Chair (with Gesine Schwan) of SDSN Germany, and as part of the executive council of the German UNESCO-Commission.

Contributors

ANDREA KOSCHINSKY
Professor Dr., Department of Physics and Earth Sciences, Jacobs University Bremen, Bremen, Germany

ANNA-KATHARINA HORNIDGE
Professor Dr. and Director, German Development Institute (DIE), University of Bonn, Bonn, Germany

BEN BOTELER
Institute for Advanced Sustainability Studies, Potsdam, Germany

CAROLE DURUSSEL
Dr., Institute for Advanced Sustainability Studies, Potsdam, Germany

EDMUND MASER
Professor Dr. and Director, Institute of Toxicology and Pharmacology for Natural Scientists, University Medical School, Kiel, Germany

JENNIFER S. STREHSE
Dr., Institute of Toxicology and Pharmacology for Natural Scientists, University Medical School, Kiel, Germany

LUISE HEINRICH
Department of Physics and Earth Sciences, Jacobs University Bremen, Bremen, Germany

MAARTEN BAVINCK
Professor Dr., Department of Geography, Planning and International Development Studies, University of Amsterdam, Amsterdam, Netherlands

MARTIN VISBECK
Professor Dr., University of Kiel, Kiel, Germany and Head of the Physical Oceanography Research Unit, GEOMAR Heimholtz Centre for Ocean Research, Kiel, Germany

SEBASTIAN UNGER
Institute for Advanced Sustainability Studies, Potsdam, Germany

SIGRID KEISER
Physical Oceanography Research Unit, GEOMAR Heimholtz Centre for Ocean Research, Kiel, Germany

TOBIAS BÜNNING
Institute of Toxicology and Pharmacology for Natural Scientists, University Medical School, Kiel, Germany

TORSTEN THIELE
Institute for Advanced Sustainability Studies, Potsdam, Germany

WERNER EKAU
Dr., Leibniz Centre for Tropical Marine Research, Bremen, Germany

Abstracts

Preface to Transitioning to Sustainable Life below Water
by Werner Ekau and Anna-Katharina Hornidge

The formulation of the Agenda 2030 of the United Nations in 2015 has given rise to a substantial increase of attention payed to the sustainable governance of our oceans. Yet, the institutional landscape remains fragmented and driving an ecologically and socially just transition of existing ocean governance practices forward remains a challenge. This introductory chapter gives an insight into the background discussions that led to this edited book on the Sustainable Development Goal 14 and offers an overview over the structure of the book.

Ocean Pollution—A Selection of Anthropogenic Implications
by Jennifer S. Strehse, Tobias H. Bünning and Edmund Maser

The number of chemicals that have a detrimental influence on the world's marine systems is almost uncountable. Some of them, e.g., mercury and its organic compounds, have to an extent always been part of global cycles, but have been released in far greater amounts since the beginning of the Industrial Revolution in the late 18th century. Others, such as pharmaceuticals, persistent organic pollutants (POPs) and microplastics, have emerged only during the last 100 years, but in alarming proportions. Their influence on marine ecosystems is often very poorly understood. Sources are as diverse as the substances are: sewage water (e.g., pharmaceuticals and microplastics), atmospheric release (e.g., metals and POPs) and even intentional dumping of waste (e.g., munitions and metals). Many of these substances can be found even in pristine areas such as the deep-sea Mariana trench or the perpetual ice of the Arctic. While problematic on their own, it seems that some of these compounds interfere with each other. Microplastics, for example, are discussed as a vector for hydrophobic organic components such as POPs, metals and microbiota. This chapter provides an overview on the topics of microplastics, persistent organic pollutants, metals, munitions, and pharmaceuticals. The condition of the Baltic Sea, which is considered the most polluted sea in the world, is given as an example. These pollutants serve as representative and well-recognized contaminants, which are gaining more public attention. It is intended to serve as an introduction to further research on ecotoxicologically relevant chemicals.

The Featuring of Small-Scale Fishers in SDG 14: Life below but Also above Water
by Maarten Bavinck

This chapter deals with SDG 14 ('life below water') and the curious inclusion of a clause (Article 14B) supporting the global population of small-scale fishworkers and their marine livelihoods. It enquires as to the background of this particular clause, which it traces back to the drawn-out international negotiation process that was marked by the ratification of FAO's Code of Conduct for Responsible Fisheries (1995) and the Voluntary Guidelines for Securing Sustainable Small-Scale Fisheries (2014). Social movements, NGOs and the academic community are demonstrated to have played an important role in both achievements. While Article 14B provides an additional impulse to the international small-scale fisheries movement, the challenges that small-scale fishworkers face are still substantial. As Article 14B concedes, the struggle focuses on maintaining access to resources and markets, but also responding to the critiques of the conservation movement.

Global Processes in Ocean Policy: An Opportunity to Create Coherence in Governance Frameworks and Support the Achievement of Conservation Goals
by Ben Boteler, Carole Durussel, Sebastian Unger, Torsten Thiele

Three major global processes in ocean governance under the umbrella of the United Nations are currently underway: negotiations for an international legally binding agreement under the 1982 United Nations Convention on the Law of the Sea (UNCLOS) for the conservation and sustainable use of marine biodiversity in areas beyond national jurisdiction (BBNJ); the 20 Aichi Biodiversity Targets adopted in 2010 as part of the Convention on Biological Diversity (CBD) Strategic Plan for Biodiversity 2011–2020 are coming to an end and new and updated biodiversity targets will be adopted as part of the Post-2020 Global Biodiversity Framework in 2020; and many of the targets set under the Sustainable Development Goal 14 (SDG 14) as part of the ocean United Nations 2030 Agenda, which includes focuses on 17 Sustainable Development Goals, to holistically address current global challenges are set to expire and are expected to be updated or renewed. This Chapter highlights the need to ensure coherence across these global processes for marine conservation and provides ways in which ocean governance can be strengthened to support global processes and marine conservation goals.

Climate Change and Its Impact on the Ocean
by Martin Visbeck and Sigrid Keiser

Increased human activities—in particular energy generation and land use—have led to atmospheric pollution by the significant emission of greenhouse gases such as carbon dioxide (CO2) and methane. The associated climate change is also affecting the ocean while, at the same time, the ocean plays a fundamental role in mitigating climate change by serving as a major heat and carbon sink. We highlight some of the most salient aspects of climate change impacting the ocean as articulated in the Special Report on the Ocean and Cryosphere in a Changing Climate by the Intergovernmental Panel on Climate Change (IPCC) released in 2019. It shows that the ocean is warming, the global sea level is rising, ocean heatwaves are more frequent, the ocean is becoming more acidic, marine ecology is shifting, levels of dissolved oxygen are reducing and the melting of ocean-terminating glaciers and ice sheets around Greenland and Antarctica is rapidly increasing. From the perspective of meeting the United Nations Sustainable Development Goals, in particular SDG 14, there are strong synergies between promoting climate mitigation and adaptation strategies, which are enshrined in SDG 13 and outlined in more detail by the Paris Agreement. Scientific research and solution-oriented knowledge generation require the growth and transformation of the science system. Specifically, they will require more freely shared ocean data, new and more effective ways of analyzing observational data fused with ocean and climate models, and enhanced timely assessment, predictions and scenario development of future ocean conditions. At the same time, knowledge from natural and social sciences, as well as informal knowledge, must be considered. Ocean science must be in a position to support decision makers by providing knowledge and frameworks to weigh the ecological, environmental and human impacts with an expected increase in use of the ocean for different sustainable development pathways. In recognition of this challenge, the United Nations declared 2021–2030 as the Decade of Ocean Science for Sustainable Development in order to advance "the science that we need for the ocean we want". The ocean decade seeks to catalyze a change towards more international, shared and solution-oriented ocean science.

Deep-Sea Mining: Can It Contribute to Sustainable Development?
by Luise Heinrich and Andrea Koschinsky

Deep-sea mining is increasingly suggested to meet the metal demand of the growing world population and to bring revenue and resource independence to many countries. Deep-sea mining is often also presented as a source for the metals required for the transition to a low-carbon economy. However, the exploitation of marine mineral resources will also be associated with considerable adverse impacts. Therefore, it is necessary to assess deep-sea mining impacts from a sustainability perspective and discuss if and how deep-sea mining could be compatible with sustainable development. Although deep-sea mining describes the extraction of a finite resource and, therefore, appears to contradict the Brundtland definition of sustainable development, this assessment finds that deep-sea mining could, under certain conditions, contribute to sustainable development. Important pre-requisites for this include the availability of an effective fiscal and revenue management system to ensure that the returns from deep-sea mining secure long-term benefits for national economies and stringent environmental regulations. Furthermore, environmental and social impact assessments have to be conducted early in the process and complemented with, inter alia, sound environmental and social management plans. As the success of these measures strongly depends on the availability of trained personnel, capacity-building initiatives need to be implemented in prior or in parallel to the establishment of deep-sea mining operations. Nevertheless, there is an urgent need to explore alternatives to deep-sea mining, including the increase of recycling rates, the substitution of critical materials and an overall change of consumer behavior.

Preface to Transitioning to Sustainable Life below Water

Anna-Katharina Hornidge and Werner Ekau

1. Introduction

The ocean has played a major role for humanity for thousands of years. It connects and separates people and nations, and provides space for trade and transport, raw materials and food. Just as old is the attempt to regulate the use of the ocean and to use it for one's own interests. It took until 1967, however, for Arvid Pardo, a diplomat from Malta, who represented his small island state, which had recently become independent, at the UN as an ambassador, to give a much-noticed, four-hour speech in which he denounced the unchecked exploitation of the seas and the seabed and the unequal distribution of opportunities, especially for small states, to share in the fruits of the sea. His call for the oceans as a whole to be declared the heritage of all humanity and for mechanisms to be developed to give all states equal opportunities to use the oceans ultimately led to the 3rd Conference on the Law of the Sea and the formulation of the United Nations Convention on the Law of the Seas (UNCLOS) in 1982. Unfortunately, UNCLOS is mainly concerned with the territorial seas as national economic zones (EEZs) and the seabed. The open ocean beyond the EEZ is not covered.

The 3rd Conference on the Law of the Sea, which consisted of 10 sessions from 1973 to 1982, also led to discussions in other UN agencies on better ocean management. However, the approaches were sectoral. Juda and Burroughs (1990) and Alexander (1993) urged a trans-sectoral approach to sustainable management.

This step was only achieved with the signing of the Convention on the implementation of the Sustainable Development Goals (SDGs) in 2015, in which the member states of the United Nations issued a universal call to action to effectively fight poverty, protect our planet and ensure that all people can live in peace and prosperity by 2030.

Finally, the largest habitat we have on this planet, the ocean, is also given recognition of the importance it has for humanity with the dedication of its own goal, SDG 14 *Life below Water*.

The oceans drive global energy and material flow systems. They are thus largely responsible for the distribution of nutrients and heat on the planet, absorbing heat from the atmosphere and storing it, thus acting as a buffer for natural fluctuations.

The oceans absorb about 30% of the carbon dioxide produced by humans. The absorption of carbon dioxide manifests as a reduction in pH, commonly referred to as ocean acidification, and affects the oxygen content of the water. This has led to an increase in oxygen depletion in many coastal waters and major upwelling areas worldwide (Zhang et al. 2010). Human-sponsored marine pollution is reaching alarming levels, with nutrient inputs, toxic chemicals, plastics and munition debris having long-lasting negative impacts on ecosystems (see the following chapters).

In this context, SDG 14 is embedded in the complex network of 17 goals in total, all of which are interrelated, interdependent and call for a move towards social, economic and environmental sustainability. The sustainable management of our marine ecosystems, coasts and marine resources will remain our greatest challenge in the 21st century.

Relatively recent shifts in policy discourse have thus turned towards developing approaches of integrated marine and coastal governance. The empirical and theoretical knowledge base, however, is substantially less developed than that with regard to terrestrial systems. Thus, sciences within the field of ocean governance are, first, increasingly concerned with building sound empirical and theoretical bases for understanding the complexities of governing coastal and marine spaces; and second, with fostering science–policy dialogues that assure close interactions and mutual transformative learning for building ocean governance frameworks and instruments on international, regional and national levels that meet the challenges of increased uses and limited carrying capacities of the ecosystems themselves. The aim of these debates is to develop governance instruments that are applicable across sectors and ecosystems and at local, national and regional levels (Schlüter et al. 2020; Kirkfeldt et al. 2021; Gissi et al. 2021).

This volume brings together a number of papers giving insight into the knowledge bases with regard to marine ecosystems and their governance challenges, as well as reflecting the policy environment and governance instruments needed for meeting the Agenda 2030 formulated by the UN in 2015 based on these insights (OceanGov 2020).

2. Climate Regulator, Biodiversity Hub and Resource Provider

The ocean drives global energy and material fluxes and acts as global climate regulator, hosts enormous biodiversity and is a key source of protein supply for humans (FAO 2020; IPCC 2019; IPBES 2019). It captures large quantities of carbon and produces around 50% of atmospheric oxygen. The absorption of carbon dioxide results in a reduction in the pH value, commonly referred to as ocean acidification, which affects the oxygen content of the water. The latter has led to an increase in oxygen depletion in many coastal waters and large upwelling areas worldwide, changing organisms' life cycles and whole ecosystems. Marine pollution, which is mainly caused by humans, has reached alarming levels, with nutrient inputs, toxic chemicals, plastics and ammunition residues having long-lasting negative effects on ecosystems. About half of humanity is directly or indirectly dependent on marine and coastal ecosystems for its quality of life. The total value of marine ecosystem services was estimated to reach USD 21 trillion, USD 11 million thereof from coastal systems (Costanza et al. 1997). The market value of marine and coastal resources and industries is estimated at USD 3 trillion per year, or about 5% of the global GDP (UNCTAD 2021). According to the UN Food and Agriculture Organization (FAO), fisheries and marine aquaculture form the basis of livelihood provision for some 12% of the world's population. In many developing countries, fish and seafood are an essential source of protein. We cannot stand by as our coasts and seas continue to be polluted and resources over-exploited, and human-driven climate change is changing entire ecosystems. Not only should we, but we must use the marine ecosystem in a way that keeps it healthy and provides us with optimum of services. Sustainability means using the sea intelligently for the benefit of all humanity. This is the basic statement behind SDG 14 *Life below water*.

The international community has, thus far, not sufficiently addressed these challenges. The strategic and economic relevance of the ocean and its resources decisively influences the negotiation processes and leads to protracted negotiations. This applies to the awarding of deep-sea mining rights via the International Seabed Authority, as well as agreements on the handling of biological resources and information from high seas areas (Biodiversity Beyond National Jurisdiction (BBNJ).

3. Ocean Governance for a Sustainable Future

As pointed out above, it is only since the late 1960s that, initiated by Arvid Pardo and supported by Elisabeth Mann Borgese in the 1970s, the idea of the ocean as the common heritage of mankind, was adopted by UN diplomacy and triggered the development of the UN Convention on the Law of the Sea (UNCLOS). However, the principle applies exclusively to the ocean floor and its mineral resources in areas beyond national jurisdiction. Coastal waters (EEZ; within 200 nm) with their living and non-living resources are under national jurisdiction; however, the high sea, with its valuable fish and biological resources, is not regulated, which leads to extensive illegal, unregulated and unreported (IUU) fishing.

In order to achieve significant progress in combating IUU, distributing resources equitably and ensuring equal rights for all, for example, international and transregional cooperation must be promoted, which allow negotiations at eye level geared towards a global common good. Geopolitical tensions are sometimes based on different value systems, for example, with regard to human rights, political regimes, or the value of ecosystems, biodiversity and a healthy ocean. It must be understood that global challenges such as those caused by climate change or the COVID-19 pandemic can only be solved jointly and in transregional dialogue.

Thus, we need structural policies that foster the global common good. The ocean here can play a key role in developing new governance models to maintain its role in stabilizing the climate, acting as home to some of the world's richest biodiversity hotspots, and being available for the upcoming blue economy. Social protection, food and health systems play important roles in contemporary crisis management and in assuring societal resilience with regard to future crises. Quality education (SDG 4) and science and innovation (SDG 9) are the core fields of action to reduce social inequalities (SDG 10), overcome poverty (SDG 1) and ensure social justice and peace (SDG 16), promote political participation, respect cultural diversity, and create a climate-neutral (SDG 13) and stable economic system, which restrict our production and consumption system in ways that CO_2 emissions are reduced, our climate stabilized and a socially just transition is assured.

The Agenda 2030 of the United Nations explicitly focuses on sustainable ocean governance (United Nations 2015). SDG 14 'Life below Water', comprising 10 targets (see annex), is dedicated to the largest habitat on the planet: a habitat that encompasses all climate zones, from the poles to the tropics and is globally connected by large, transregional currents that transport heat and nutrients. SDG 14 focusses on the protection of marine and coastal ecosystems from pollution in a sustainable way and addressing the effects of ocean acidification, the establishment of Marine Protected Areas and fighting against IUU. The formation of scientific knowledge and transfer of marine technology is thematized in the same way as the support of artisanal fisheries is requested. These targets also continue to serve as priority directions during the UN Ocean Decade 2021–2030. The SDGs have stimulated and are the basis for further discussions in other dialogues, such as the Conference of the Parties to the United Nations Framework Convention on Climate Change (UNFCCC); they have also influenced debates on the primacy of Blue Carbon and the sustainable use of marine biodiversity in the Areas Beyond National Jurisdiction (ABNJ), as well as the development of international exploitation rules for deep-sea minerals and the ambitions of some nations to extend their territories towards the open ocean.

4. Overview of the Book

In the following chapters, a number of selected and crucial threats to the ocean's ecosystems are discussed, which relate to the ten targets of SDG 14 'Life below Water' (see Appendix A): ocean warming and acidification caused by the still increasing release of CO_2; pollution as a major human threat to the ocean by releasing chemicals, nutrients and plastics into the sea and the dumping of munitions; responsible management of small-scale fisheries; and deep-sea mining. In addressing these challenges, political will and action are key. Thus, the following chapters each reflect the respective ecosystem challenge from a governance perspective.

As part of the ongoing efforts of Agenda 2030 implementation as well as in preparation of the upcoming Sustainable Development Goal Summit in September 2023 of the United Nations, the volume recommends development policy-makers and researchers—and in line with Hornidge (2020)—to turn their attention to the governance of the world's ocean in the following areas:

Food system of the future: Artisanal fisheries represent the largest group of people working in capture fisheries. It is thus an important economic sector, especially on tropical coasts. Increasingly, this sector is coming under pressure as competition in coastal waters has steadily been increasing for years due to the sale of licenses to foreign industrial fishermen. Too many vessels, oversized nets, as well as too many fishermen in artisanal fishing, who are operating ever-larger motorized boats, are leading to overfishing of the stocks. The social impacts are enormous and mostly neglected by policy makers. There is a lack of alternative income opportunities for the fishermen. The only remedy is the stricter application of existing principles for sustainable fishery management, in combination with the principles of good governance and the rule of law, as formulated in the FAO Guidelines for Small-scale Fisheries (FAO 2015).

Living with coastal change processes: Coastal societies have to adapt to the multiple consequences of changes in socio-ecological systems that are taking place worldwide at an increasing pace. The intensive use of the coasts generates an increasing diversity of stakeholders (e.g., fish farmers, fishermen, tourism companies, infrastructure operators, poor and rich, regulators and regulated) and leads to conflicts of interest and power asymmetries. Here, transformative approaches need to be found that lead to sustainable cooperation between these groups, including scientists. Tools such as marine spatial planning, ecosystem-based fishery management, which includes marine protected areas and adaptations to sea-level rise and coastal erosion, require societal, technological and nature-based solutions, e.g., carbon-capturing mangrove forests.

Knowledge and knowledge partnerships for sustainable ocean governance: An essential task for further improving ocean governance is to better disseminate existing knowledge about the ocean and make it accessible. This will strengthen the negotiating position of coastal states in regional and multilateral debates on ecosystem conservation and blue economy job creation. The SDGs, including the Summit 2023 and the United Nations Decade of Ocean Science for Sustainable Development (2021–2030), are important tools to build and expand necessary regional networks between decision-makers and scientists in the field of ocean governance. The formulation and implementation of sustainability standards (ecological, social, economic and cultural) on a local and regional level and the consistent further development of the "Blue Economy" principles must be one of the goals for the next decade.

The ocean, as the largest contiguous habitat, a source of food, a service provider for the exchange of goods and a buffer for climate fluctuations has finally become prominent in the global political agenda. We must seize the opportunity to set the right path for a sustainable, shared future of our planet.

Author Contributions: Conceptualization, writing—original draft preparation, review and editing were done in close cooperation between the authors A.-K.H. and W.E. All authors have read and agreed to the published version of the manuscript.

Funding: This research received no external funding.

Acknowledgments: The presented work rests on discussions led by the European COST-Action Network of researchers and policy-makers entitled "Ocean Governance for Sustainability—challenges, options and the role of science", CA15217 (www.oceangov.eu; accessed on 21 February 2022). We would like to thank COST for the funding that made the cooperation amongst the authors, and thus, this article, possible. Further, we thank the publisher for the opportunity to publish this volume on SDG 14 and hope that it will be widely distributed.

Conflicts of Interest: The authors declare no conflict of interest.

Appendix A. Sustainable Development Goal 14—United Nations 2015

Target 14.1: By 2025, prevent and significantly reduce marine pollution of all kinds, in particular from land-based activities, including marine debris and nutrient pollution.

Indicator: Index of (a) coastal eutrophication and (b) floating plastic debris density.

Target 14.2: By 2020, sustainably manage and protect marine and coastal ecosystems to avoid significant adverse impacts, including by strengthening their resilience, and take action for their restoration in order to achieve healthy and productive oceans.

Indicator: Number of countries using ecosystem-based approaches to managing marine area.

Target 14.3: Minimize and address the impacts of ocean acidification, including through enhanced scientific cooperation at all levels.

Indicator: Average marine acidity (pH) measured at agreed suite of representative sampling stations.

Target 14.4: By 2020, effectively regulate harvesting and end overfishing, illegal, unreported and unregulated fishing and destructive fishing practices and implement science-based management plans, in order to restore fish stocks in the shortest time feasible, at least to levels that can produce maximum sustainable yield as determined by their biological characteristics.

Indicator: Proportion of fish stocks within biologically sustainable levels.

Target 14.5: By 2020, conserve at least 10% of coastal and marine areas, consistent with national and international law and based on the best available scientific information.

Indicator: Coverage of protected areas in relation to marine areas.

Target 14.6: By 2020, prohibit certain forms of fishery subsidies which contribute to overcapacity and overfishing, eliminate subsidies that contribute to illegal, unreported and unregulated fishing and refrain from introducing new such subsidies, recognizing that appropriate and effective special and differential treatment for developing and least developed countries should be an integral part of the World Trade Organization fisheries subsidies negotiation.

Indicator: Degree of implementation of international instruments aiming to combat illegal, unreported and unregulated fishing.

Target 14.7: By 2030, increase the economic benefits to developing small-island states and the least economically developed countries from the sustainable use of marine resources, including through the sustainable management of fisheries, aquaculture and tourism.

Indicator: Sustainable fisheries as a proportion of GDP in developing small-island states, the least economically developed countries, and all countries worldwide.

Target 14.a: Increase scientific knowledge, develop research capacity and transfer marine technology, taking into account the Intergovernmental Oceanographic Commission Criteria and Guidelines on the Transfer of Marine Technology, in order to improve ocean health and to enhance the contribution of marine biodiversity to international development, in particular in developing small-island states and the least economically developed countries.

Indicator: Proportion of total research budget allocated to research in the field of marine technology.

Target 14.b: Provide access for small-scale artisanal fishers to marine resources and markets.

Indicator: Degree of application of a legal/regulatory/policy/institutional framework which recognizes and protects access rights for small-scale fisheries.

Target 14.c: Enhance the conservation and sustainable use of oceans and their resources by implementing international law as reflected in UNCLOS, which provides the legal framework for the conservation and sustainable use of oceans and their resources, as recalled in paragraph 158 of "The Future We Want".

Indicator: Number of countries making progress in ratifying, accepting and implementing ocean-related instruments through legal, policy and institutional frameworks, which implement international law, as reflected in the United Nation Convention on the Law of the Sea, for the conservation and sustainable use of the oceans and their resources.

References

Alexander, Lewis. 1993. Large marine ecosystems: A new focus for marine resources management. *Marine Policy* 17: 186–98. [CrossRef]

Costanza, Robert, Ralph dArge, Rudolf de Groot, Stephen Farber, Monica Grassot, Bruce Hannon, Karin Limburg, Shahid Naeem, Robert V. O'Neill, Jose Paruelo, and et al. 1997. The value of the world's ecosystem services and natural capital. *Nature* 387: 253–60. [CrossRef]

FAO. 2015. *Voluntary Guidelines for Securing Sustainable Small-Scale Fisheries in the Context of Food Security and Poverty Eradication*. Rome: FAO, Available online: http://www.fao.org/3/i4356en/I4356EN.pdf (accessed on 22 February 2022).

FAO. 2020. *The State of World Fisheries and Aquaculture 2020*. Rome: FAO, Available online: http://www.fao.org/state-of-fisheries-aquaculture (accessed on 9 September 2020).

Gissi, Elena, Frank Maes, Zacharoula Kyriazi, Ana Ruiz-Frau, Catarina Frazao-Santos, Barbara Neumann, Adriano Quintela, Fatima L. Alves, Simone Borg, Wenting Chen, and et al. 2021. Contributions of marine area-based management tools to the UN sustainable development goals. *Journal of Cleaner Production* 330: 129910. [CrossRef]

Hornidge, Anna-Katharina. 2020. The Ocean as a Lifeline for the Future of the Planet, Bonn: German Development Institute/Deutsches Institut für Entwicklungspolitik (DIE), The Current Column of 8 June 2020. Available online: https://www.die-gdi.de/en/the-current-column/article/the-ocean-as-a-lifeline-for-the-future-of-the-planet/ (accessed on 22 February 2022).

IPBES (Intergovernmental Science-Policy Platform on Biodiversity and Ecosystem Services). 2019. Global Assessment Report on Biodiversity and Ecosystem Services. Available online: https://ipbes.net/global-assessment (accessed on 22 February 2022).

IPCC. 2019. Special Report on the Ocean and Cryosphere in a Changing Climate. Available online: https://www.ipcc.ch/srocc (accessed on 22 February 2022).

Juda, Lawrence, and Richard Burroughs. 1990. The prospects for comprehensive ocean management. *Marine Policy* 14: 23–35. [CrossRef]

Kirkfeldt, Trine Skovgaard, Jan P. M. van Tatenhove, and Helena M. G. P. Calado. 2021. The Way Forward on Ecosystem-Based Marine Spatial Planning in the EU. *Coastal Management* 2021: 1–16. [CrossRef]

OceanGov. 2020. Ocean Governance for Sustainability, COST Action Website. Available online: www.oceangov.eu (accessed on 22 February 2022).

Schlüter, Achim, Kristof van Assche, Anna-Katharina Hornidge, and Natasa Vaidianu, eds. 2020. Land-sea interactions and coastal development: An evolutionary governance perspective. *Special Issue in Marine Policy* 112: 103801. [CrossRef]

UNCTAD. 2021. UNCTAD e-Handbook of Statistics 2021. Available online: https://hbs.unctad. org/world-seaborne-trade/ (accessed on 1 February 2022).

United Nations. 2015. *Transforming our world: The 2030 Agenda for Sustainable Development.* New York: United Nations, pp. 1–41.

Zhang, Jing, Denis Gilbert, Andrew Gooday, Lisa A. Levin, S. Wajih A. Naqvi, Jack Middelburg, Mary I. Scranton, Werner Ekau, Angelica Pena, Boris Dewitte, and et al. 2010. Natural and human-induced hypoxia and consequences for coastal areas: synthesis and future development. *Biogeosciences* 7: 1443–67. [CrossRef]

Ocean Pollution—A Selection of Anthropogenic Implications

Jennifer S. Strehse, Tobias H. Bünning and Edmund Maser

1. Microplastics in the Marine Environment

1.1. Introduction—How Plastics Enter the Environment

Today, plastic is one of the most used polymers in the world and comprises one of the five common material classification categories alongside metallic, ceramic, organic and composite materials. From its creation in the 1870s, it has become an integral part of human life, mainly due to its advantageous properties relating to elasticity, lightness, versatility and durability. It is clear, meanwhile, that plastic has profoundly changed daily life. Since the 1940s, annual plastic production has massively increased over the decades from 0.5 million tons to nearly 370 million tons in 2019 (Plastics Europe 2020).

With regard to the molecular structure, about 80% of the polymers in Europe include polypropylene (PP), polyethylene (HD-PE and LD-PE), polyvinyl chloride (PVC), polyurethane (PUR), polyethylene terephthalate (PET) and polystyrene (PS) (Plastics Europe n.d., Annual Review 2017–2018).

Applications of plastics are innumerable and invaluable; however, the widespread use of synthetic polymers on the one side, together with their high durability on the other, is a great disadvantage with regard to its persistence in the environment. Plastics, when introduced—accidentally or deliberately—into the environment, pose a long-term and increasing challenge and threat to the environment, including the marine biota and, lastly, human population (Wright and Kelly 2017; Rist et al. 2018). Unfortunately, recycling of plastics is performed only to a very small extent. For example, in 2013 only 14% of the total mass of plastic packaging materials was recycled, whereas a bulk of around 72% was either dumped in landfills or released into the marine environment (World Economic Forum 2016). However, besides landfills, a large proportion of plastic waste is currently incinerated and used for energy production.

In 2014, the United Nations Environment Program (UNEP) identified plastic pollution in the oceans, and its consequences for the marine ecosystem, as one of the top ten emerging global environmental problems (UNEP 2014).

On average, more than eight million metric tons of plastics are dumped into oceans every year (Jambeck et al. 2015; UNEP 2017), leading to spectacular examples of plastic litter accumulation in the oceans, for example, the Pacific garbage patch, North Atlantic garbage patch (Atlantic Ocean), and Indian Ocean garbage patch (Lebreton et al. 2018; Dąbrowska et al. 2021). Approximately five trillion tons of plastic debris are estimated to be floating in the oceans around the globe (Eriksen et al. 2014; Barboza et al. 2018). The severity of plastic pollution is highlighted by reports of plastics in previously pristine marine waters such as the Antarctic, Arctic and the deepest point on earth (Mariana Trench) (Mendoza et al. 2018; Ross et al. 2021). However, the estimate that there will be more plastics in the oceans (by weight) than fish by the end of the year 2050 is false or at least uncertain, because there is uncertainty over the methods used to estimate both fish populations and the amount of plastic in the oceans (GRID-Arendal 2021).

While considerable amounts of the plastics end up in marine ecosystems, they affect marine organisms via entanglement and/or ingestion. Entanglement in plastic litter has been reported for a wide variety of organisms including mammals and cetaceans (Laist 1997; Gall and Thompson 2015; Uddin et al. 2020; Khalid et al. 2021). In addition to all kinds of drifting plastic debris, fragments of discarded or lost fishing nets are of particular danger. Once entangled, organisms suffer from reduced mobility and feeding, or at worst can drown, suffocate and become strangulated (Li et al. 2016).

Ingestion of plastic particles, mistakenly considered as food, occurs in most marine organisms ranging from plankton to fish, birds and turtles. When ingested, plastics cannot only lead to irritation and injuries in the digestive tracts of organisms, but can also result in a false sensation of satiation, impacting the fitness and reproduction of marine organisms (GESAMP 2016; Andrades et al. 2019; Zantis et al. 2021; López-Martínez et al. 2021; Akindele and Alimba 2021).

1.2. Primary and Secondary Microplastics

Plastics are very resistant to decomposition and may stay in the environment for centuries (Rhodes 2018). Several processes of embrittlement and weathering at sea, including UV-B-induced photo-oxidation, hydrolysis, thermal degradation, biodegradation and mechanical erosion lead to plastic fragmentation. Additionally, mechanical properties such as strength, hardness, ductility and stiffness can vary significantly among different types of plastics. During the slow process of decomposition, large plastic particles eventually break down to meso-, micro- and nanoplastics, commonly known as secondary microplastics.

By definition, marine plastic debris is divided into four size categories: macroplastics (>20 mm), mesoplastics (5–20 mm), microplastics (<5 mm) and nanoplastics (<100 nm) (Frias and Nash 2019; Hartmann et al. 2019; Hidalgo-Ruz et al. 2012). Sometimes the term mega-plastics (>100 mm) is used. Eventually, the National Oceanic and Atmospheric Administration (NOAA, Silver Spring, MD, USA) and the European Marine Strategy Framework Directive (Directive 2008/56 2008) Technical Subgroup on Marine Litter proposed that plastic debris < 5 mm be considered microplastics; this definition is most frequently used. Despite this consensus, it would be more appropriate to consider plastic particles smaller than 1 mm as micro-plastics, as the prefix micro refers to the micrometer range (Van Cauwenberghe et al. 2015). Even more, in order to avoid a gap in size between micro- and nanoplastics, a meaningful size definition for microplastics would be 100 nm to 5 mm, as nanoplastics are defined by the International Union of Pure and Applied Chemistry (IUPAC) as smaller than 100 nm (Vert et al. 2012). However, the definition of primary and secondary microplastics is still debated, as the boundaries are not always very clear or useful. For instance, tire particles and microfibers are both derived from fragmentation and wear-and-tear of larger objects, similar to most secondary microplastics. It should be noted further that different terms are sometimes used, e.g., fragments, pellets, beads, granules, spherules, discs, fibers, filaments, plastic films, foamed plastic, Styrofoam, etc. (Hidalgo-Ruz et al. 2012; Frias and Nash 2019; Hartmann et al. 2019).

A second type of microplastics called primary microplastics is directly manufactured in <5 mm sizes and released accidentally into the environment. These particles include microfibers used in textiles, microbeads or micro-pellets used in facial cleansers, toothpaste and cosmetics, industrial scrubbers used for abrasive blast cleaning and capsules for drug delivery (Ivleva et al. 2017). In addition to being produced as intended products, wastes from manufacturing processes or derivatives from erosion and tearing in the use of large plastic products, such as tires, wheels, boards, etc., belong to this type of microplastics (Sieber et al. 2020).

Primary microplastics resulting from personal cleaning products are introduced into the oceans via sewage effluents (Mintenig et al. 2017; Gago et al. 2018). Synthetic fibers are released from clothing due to mechanical forces acting in washing machines (Imhof et al. 2016; De Falco et al. 2018). These findings underline the importance of household inputs. Moreover, considerable volumes of sewage sludge and effluent are discarded into the sea. For example, the concentration of microplastic particles in sludge was recorded in the range from 6.7×10^3 to 62.6×10^3 particles per 1 kg dry matter (Eckert et al. 2018). Virgin pellets used in the polymer industry enter the environment due to accidental or deliberate spillage from factories, or during transport (Hidalgo-Ruz et al. 2012). Additionally, wind, storm events, soil erosion and run-offs from drainage basins play an important role in the entry of microplastics from land-based sources into aquatic systems (Vianello et al. 2013). It has been estimated that 1.5 million tons of primary microplastics are released into water yearly, and that microplastics have already been ubiquitously reported in almost all aquatic habitats of the planet, from the open seas to deep oceans, river, lakes, the water column, and sediments (Picó and Barceló 2019).

Marine debris originating from land were estimated at 80%, including riverine and direct inputs (Mani et al. 2016). The remainder was attributed to shipping such as industrial, fishing and recreational boating activities (Claessens et al. 2011). In aquatic environments, microplastics can be located at the surface, in the water column and in bottom or beach sediments. Even deep sea sediments have been found to serve as an ultimate repository for microplastics (Woodall et al. 2014; Cunningham et al. 2020).

Factors determining the distribution of microplastics in the oceans are enormously complex and make it difficult to establish precise quantitative modeling and meaningful data correlation (Van Sebille et al. 2020). Horizontal transport is presumably dependent on currents and waves, while vertical transport is probably influenced by temperature but also, of course, by densities, sizes and shapes of the particles. It appears plausible that the highest concentrations of microplastics are present in coastal regions or in regions with high anthropogenic activities, i.e., industrial, shipping, fishing or touristic.

Taken together, it is clear today that microplastic occurrence is a globally omnipresent phenomenon. However, effects are still largely unknown but are increasingly the subject of scientific scrutiny, as microplastic pollution is suspected to rapidly increase in water bodies in the future (Wright and Kelly 2017; Rist et al. 2018). Moreover, even smaller plastic particles, the so-called nanoplastics, represent a potentially hazardous material as well, especially due to their smaller sizes and unpredictable impacts at cellular level.

1.3. The Impact of Microplastics on Marine Biota

Importantly, due to their small size, micro- and nanoplastics are of particular concern as: (1) they can move fast and far in the environment; (2) they have a relatively large surface for sorption of pollutants and constituent chemicals; (3) they may contain additives, such as plasticizers, inorganic fillers, thermal and UV stabilizers, fire retardants and colorants, etc.; (4) they can migrate through tissues of animals; (5) they can easily enter the food web and concentrate along trophic levels up to human beings (Huang et al. 2020).

The extent of deleterious physical impact on biota caused by microplastics is currently not exactly known (Rochman et al. 2016; Bucci et al. 2020). Microplastics can be regarded as a size equivalent to lower trophic level biota as macroplastics are to higher biota; therefore, the ingestion potential of microplastics must be considered. Higher trophic planktivores might mistake microplastics for prey or passively ingest them with prey, due to their high resemblance to planktonic organisms. Factors that affect ingestion of microplastics by lower trophic level biota are density, size, shape, color and abundance, making its role as a contaminant in the marine food web highly complex.

For example, microplastics (fluorescent polystyrene beads) are taken up by zooplankton, i.e., different copepod species (Cole et al. 2013; Kokalj et al. 2018; Wang et al. 2019a). Filter feeding organisms, detritivores and planktivores ingest microplastics, because they appear in the same size range as sediment particles and certain planktonic organisms. Once absorbed microplastics can impact the health of marine organisms, causing internal abrasions, ulcers, gut blockages and starvation; factors that may ultimately lead to an imbalance in the whole ecosystem by affecting some species more than others (Cappello et al. 2021; López-Martínez et al. 2021; Missawi et al. 2021; Teng et al. 2021).

Thus far, the issue of systemic absorption of microplastics by marine organisms has been controversially discussed. Whereas some studies reported the accumulation of microplastics in the hemolymph and hemocytes of mussels (Browne et al. 2008; Magni et al. 2018; Cappello et al. 2021), it has indeed been shown that blue mussels took up microplastics into their cells and tissues, causing a significant decrease in lysosomal membrane stability and formation of granulocytoma (Moos et al. 2012).

De Witte et al. (De Witte et al. 2014) reported the uptake of microplastic fibers by wild mussels from Belgian retailers and the Belgian coast, respectively, linking the problem of microplastic contamination directly to human consumers.

It can be assumed, on the basis of uptake by lower trophic organisms, that transfer of microplastics to the marine food chain and subsequent biomagnification may occur (Huang et al. 2020). The occurrence of microplastics in fish has already been observed in different demersal, mesopelagic and pelagic fish species (Lusher et al. 2013), providing a possible entry path to higher trophic levels of the food chain, such as tunas, squid and other predators (Boerger et al. 2010; Romeo et al. 2015; Chagnon et al. 2018; Oliveira et al. 2020; Zhang et al. 2021).

1.4. Microplastics as Vectors for Hydrophobic Organic Compounds, Metals and Microbiota

As a matter of fact, microplastics can absorb surrounding persistent organic pollutants (POPs) and/or heavy metals, and serve as attachment media for microorganisms. Thus, microplastics may act as vectors for lipophilic toxic chemicals and microbial pathogens to organisms (Koelmans et al. 2016; Besseling et al. 2019; Hildebrandt et al. 2020; Nobre et al. 2020; Qiu et al. 2020; Sharma et al. 2020). Of similar relevance is that microplastics release toxic additives contained in almost all plastic materials. Once degraded enough, microplastics can also release plastic monomers, thereby gaining another level of environmental relevance. More than 50% of plastics are associated with hazardous monomers, additives and chemical byproducts (Lithner et al. 2011).

For example, Mato et al. (2001) found microplastics to strongly accumulate polychlorinated biphenyls (PCBs) and dichlorodiphenyldichloroethylene (DDE), with concentrations elevated by 10^5–10^6 in the surrounding seawater (2001). These findings result from the high hydrophobicity of the compounds, which is explained by high octanol/water partition coefficients (Log K_{OW}). Especially values for PCBs and polycyclic aromatic hydrocarbons (PAHs) are alarmingly high, since dioxin-like PCBs and different PAHs are toxic chemicals that impact the ecosystem and human health. It is worth noting that these compounds have been classified by the International Agency for Research on Cancer (IARC) as Group 1 carcinogens. It has, therefore, been hypothesized that ingestion of microplastics by aquatic biota increases the bioaccumulation of anthropogenic pollutants (Nobre et al. 2020; Qiu et al. 2020; Sharma et al. 2020; Pandey et al. 2021). Microplastics may further serve as transport vehicles for POPs between different contamination areas (Wang et al. 2020).

In this context, the relevance of metals sorbed to plastic particles is still poorly understood and appears negligible compared to POPs. Nevertheless, accumulation of metal occurs. Microplastics were exposed to different metals in harbor sea water for eight weeks resulting in the following order of metal accumulation: Fe > Al > Mn > Pb > Cu,Zn > Ag (Ashton et al. 2010).

Microplastics also constitute a possible vehicle for transport of microorganisms, including potentially pathogenic microorganisms such as *Vibrio* spp. (Kirstein et al. 2016). In general, this transporting mechanism might introduce alien species into ecosystems or influence whole populations by serving as a hard-structure habitat for rafting communities and as an oviposition resource (Naik et al. 2019; Bowley et al. 2021).

1.5. Microplastics and Climate Change

For future scenarios, an interesting question is whether the expected global climate change will intensify the negative effects of microplastics on life in the oceans. This is an emerging issue with only few reports to date. Hiltunen et al. (2021) studied effects of decreasing food quality, temperature increase and microplastic exposure on the model freshwater cladoceran *Daphnia magna* and did not find any impact of microplastics on survival, size nor reproduction. Likewise, an increase in water temperature significantly affected the activity, energy reserves, oxidative stress and immune function of the freshwater mussel *Dreissena polymorpha*, while, in contrast, the effects by microplastics were limited to a change in the antioxidative capacity without any interactive effects between microplastics and thermal exposure (Weber et al. 2020).

However, since the ocean is the largest active carbon pool on the planet and plays an important role in global climate change, marine plastics may impact the gas exchange and circulation of marine CO_2, thus causing more greenhouse gas emissions. This aspect has recently been discussed by Shen et al. (2020), who hypothesized that marine microplastics affect photosysthesis, growth, development and reproduction of phyto- and zooplankton, respectively, thereby affecting the ocean carbon stock and contributing to global warming.

1.6. Outlook

Aside from the occurrence of microplastics per se, one current problem is the difficulty to establish the distribution and to quantify the amount of microplastics in waters which is attributed to the lack of proper and harmonized sampling and analysis methods (Bordós et al. 2021; Kirstein et al. 2021). This concern is parallel to the increase in introduction of microplastics into the marine environment, and the fact that microplastics are very stable and stay in the environment long after they are discarded.

Despite the fact that there has been rapid development of research on microplastics in the last 15 to 20 years, it is imperative to find approaches to prevent water pollution by microplastics today. In more detail, there is a critical need for standardization of sampling and detection techniques, i.e., the development of standard operation procedures (SOPs), subsuming sampling, separation, purification and detection, to warrant inter-study comparability of findings (Bordós et al. 2021; Kirstein et al. 2021). This would help to quantify and trace back microplastics occurrence and determine its impact on biota and habitats. In addition, approaches to remove microplastics from water sources, including biotechnology and engineering tools are to be urgently developed. Finally, reliable scientific findings may strengthen the political impetus and drive the industry to take responsibility for contributing to the resolution of the microplastic problem.

2. Persistent Organic Pollutants (POPs)

2.1. Introduction

In 2001, after long negotiations under the auspices of the United Nation Environmental Programme (UNEP), an international agreement was adopted regarding the tremendous effects of persistent organic pollutants on the environment and nature. The resulting Stockholm Convention was effective from May 2004 (UNEP 2018a). The subject of this agreement was the ambitious goal to eliminate twelve halogenated compounds, the so-called "dirty dozen", which have been identified as extremely harmful to the environment (UNEP 2018a).

The term "persistent organic pollutants"—commonly called POPs, describes different groups of halogenated compounds. Some of them can emerge from natural incidents, e.g., during volcanic eruptions or forest fires (El-Shahawi et al. 2010), but their main origin is anthropogenic release. Most of the POPs are chlorinated or brominated aromatics, for example, polybrominated diphenyl ethers (PBDE) such as the flame-retardant decabromodiphenyl ether (c-deca-BDE), PCBs, organochlorine pesticides such as dichlorodiphenyltrichloro-ethane (DDT) and lindane as well as polychlorinated dibenzo-p-dioxins and -furans (PCDD/Fs) whose best-known representative is 2,3,7,8-tetrachlorodibenzo-p-dioxin (TCDD) (Jones and de Voogt 1999; Harrad 2010). In the Stockholm Convention, POPs from anthropogenic origin were divided into three fundamental groups: pesticides, industrial chemicals and unintentional products (byproducts) of the chemical industry. As of 2019, 24 compounds and groups were listed to be eliminated from industrial use, seven as unwanted byproducts (producers have to take care that their release is minimized), and two belong to a category of compounds whose use is only permitted under restricted conditions (UNEP 2018a). One of them is DDT, an insecticide used on a larger scale in the fight against malaria but which, on the other hand, was also responsible for the near extinction of many fish-feeding birds during the 1950s (Vitousek et al. 1997). Although of little acute toxicity, DDT and its metabolites such as dichlorodiphenyldichloroethylene (DDE) and –ethane (DDD) alter the calcium homeostasis in birds, resulting in thin, fragile eggshells (Helander et al. 1982; Fry 1995). The near extinction of its most famous symbol of freedom, the bald eagle, made USA one of the first states to ban the use of DDT by 1972 (Ehrlich et al. 1988).

From the beginning of the last century until today, several million tons of POPs have been produced, including over 1.3 million tons of PCBs alone (Jamieson et al. 2017). POPs have been released into the environment: intentionally, by waste dumping such as with the pesticides DDT and lindane; accidentally, such as with 2,3,7,8-TCDD in the Seveso disaster (Bertazzi and Domenico 1994); as byproducts of insufficiently produced chemicals (e.g., 2,3,7,8-TCDD in the herbicide "Agent Orange" during the Vietnam war) (Stellmann et al. 2003); and from contaminated waste disposal and burning (Lammel et al. 2013). They are semi-volatile, hydrophobic and fat soluble. As the name implies, these substances have extremely long biological and ecological half-life due to their robustness against UV-light and biodegradation. They have the potential to be distributed to remote areas (Hung et al. 2016) and, most importantly, show harmful effects in humans and biota (Harrad 2010; Vassilopoulou et al. 2017; Landrigan et al. 2020).

The effects of persistent organic pollutants on marine organisms are as manifold as the group itself. Their role as endocrine disruptors has been studied in many different marine species. Some examples include the influence on sexual hormones of zebrafish (0.08–0.8 ng/day TCDD; (Hutz et al. 2006)), feminization of male individuals of goby (2–30 pg/L TCDD; (Wu et al. 2001)), inhibition of growth and development of mummichog larvae (80–1250 pg/g wet weight (w.w.) PCB-126; (Rigaud et al. 2013)), weight loss and lower metabolism in the European eel (7 pg/g bodyweight PCB-126; (Van Ginneken et al. 2009)), reduced level of thyroid hormone in salmon (>1 µg/L aroclor 1254; (Lerner et al. 2007)), lower survival rates of juvenile rainbow trout to pathogenic bacteria (microinjection of 0.4 to 2 µg/egg clophen A50; (Ekman et al. 2004)) and reduced sperm counts in male guppies exposed to food dosing (10 µg/mg DDE; (Kinnberg and Toft 2003)).

High-lipophilic POPs, such as PCB-153, biomagnify through piscivorous food webs. Kelly et al. (2007) found that compounds with a medium lipophilicity but high octanol-air distribution coefficient accumulate in marine mammalian food webs, but not in the piscivorous food web. For β-HCH, a polychlorinated hexane and byproduct of the production of the insecticide Lindane, a biomagnification factor below 1 (no magnification) was found for predatory fish, but 45 (strong magnification) for marine mammals. In comparison, for PCB-153 a biomagnification factor of 7.7 (predatory fish) and 45 (marine mammals) was determined (Kelly et al. 2007). Concentrations of 70–1300 µg/g l.w. of DDT and 260-1500 µg/g l.w. of PCB resulted in eggshell thinning and reproductive failure of white-tailed eagles (Helander et al. 1982, 2002; Sonne et al. 2020).

For marine mammals such as Baltic grey seals, high concentrations of PCBs and DDTs affect the reproductive system. Concentrations of 100 µg/g lipid weight (l.w.) caused elevated cortisol production, causing hyperplasia of adrenal glands, reduced bone density and skin changes, as well as changes in the female reproductive system, e.g., occlusions, stenosis and tumors of the uterus (known as the Baltic seal disease complex) (Letcher et al. 2010; Sonne et al. 2020). They also act as endocrine disruptors for cetaceans, interfering with steroid hormones (Hoydal et al. 2017). Blubber samples from bottlenose dolphins indicate an association between high concentrations of DDT and its metabolites, and impaired testosterone homeostasis (Galligan et al. 2019).

Today, some forty years after the ban of DDT in many EU countries (Macgregor et al. 2010), and more than a decade and a half after the Stockholm Convention came into force (UNEP 2018a), persistent organic pollutants can still be detected in numerous marine organisms worldwide—albeit in decreasing concentrations. Even in pristine areas such as the Arctic (Letcher et al. 2010; Ma et al. 2016) and the Hadal zone (6000–11,000 m), the deepest zone of the ocean, considerable amounts have been detected. Jamieson et al. (2017) evaluated the number of PCBs and PBDEs in amphipod crabs in two deep sea trenches (Mariana and Kermadec) in the North and South Pacific at depths down to 10,025 m. The mean values of 382.28 ng/g dry weight (d.w.), 25.24 ng/g d.w. of PCBs, and the lower but still detectable levels of PBDEs (5.82–28.93 ng/g d.w. In Mariana and 13.75–31.02 ng/g d.w. In Kermadec) were found. Compared to crabs from one of the most polluted rivers in China, the Liaohe River, with PCB concentrations between 3.61 and 5.48 ng/g d.w. (Teng et al. 2013), the amounts measured in the Mariana trench were almost a hundred times higher (Jamieson et al. 2017). This also impressively displays the global transport capabilities of the oceans.

2.2. POPs in the Baltic Sea

POPs are one of the groups under investigation by the Baltic Marine Environment Protection Commission (HELCOM). For the Baltic Sea, a region often referred to be one of the most polluted seas (Rheinheimer 1998), concentrations of different sorts of POPs are monitored in biota (HELCOM 2010). Layer-by-layer analysis of sediment cores collected from costal and offshore areas of the Baltic sea are used to trace the course of POP contamination. While developing relatively slowly during the first half of the twentieth century, a massive increase in concentration was found in layers between the 1960s and 1980s, resulting in a total of up to 28 ng/g dry weight of eight PCBs (Sobek et al. 2015). The highest levels were found in the more southern spots, which correlated with distance to the source of emissions. Since the 1990s, following their phase out, there has been a clear trend of decreasing concentrations. For samples taken offshore, there was an observed delay of about ten years in the decrease in PCB concentrations. However, for hexachlorobenzene, even a slight increase in concentration starting in 2000 was found in these samples. From the data gained, half-lives of PCBs and PCDD/Fs in sediments of the Baltic were calculated to be 14 ± 5 years near the coast and 29 ± 15 years in offshore regions. (Sobek et al. 2015).

For Baltic grey seals, whose population had been heavily reduced by effects of the Baltic seal disease complex on their female reproductive system during the 1970s, e.g., leiomyoma, stenosis and occlusions of the uterus (Desforges et al. 2016), severely decreasing trends for PCB and DDT levels have been found in liver samples collected between 1981 and 2015 (Schmidt et al. 2020). The same was observed for mean total PCBs and DDTs in the blubber of juvenile grey seals, which decreased from 110 mg/kg l.w. (PCBs) and 192 mg/kg l.w. (DDTs) in 1968 to 15 and 2.8 mg/kg l.w., respectively, in 2010. This correlates with a 100% decrease in uterine obstruction and leiomyoma, since 2000, and 100% increased pregnancy frequency (Roos et al. 2012). Similar decrease in POP concentrations has been found in white-tailed eagle eggs and sea otter muscle tissue (Roos et al. 2012). In herrings caught from the Gulf of Bothnia, PCBs and PCDD/Fs reduced by 82–86% between 1978 and 2009, from >50 to <10 pg/g fresh weight for individuals ≥ 5 years, and from 15 to 2 pg/g f.w. for younger fish. However, 45% of the fish caught in 2009 still contained more than the maximum EU allowable concentration of PCDD/Fs (3.5 pg/g f.w. toxicity equivalents) and 36% more than that allowed for PCDD/Fs and PDBs (6.5 pg/g f.w.) (Airaksinen et al. 2014). Nyberg et al. (2015) found a decrease of 60–80% for CB-153 since 1988, 90% for DDE since the late 1970s, and 90% for HCHs and HCBs since 1979, e.g., in herring, guillemot and blue mussel. CB-118 and DDE, however, still exceeded the OSPAR target values, with 24 ng/g lipid and 5 ng/g wet weight, respectively.

While the concentrations of the heavily regulated or banned POPs such as PCBs, PCDD/Fs and DDTs steadily declined, new organohalogen compounds, such as organophosphate esters (OPEs), halogenated flame retardants (HFRs) and chlorinated paraffins (CPs) were recently found in different species from the Baltic sea (De Wit et al. 2020). Some of them have biomagnification potentials comparable to p,p-DDE and CB-153. As toxicity data are still lacking, monitoring these substances is necessary (De Wit et al. 2020).

2.3. Climate Change Might Cause Re-Emission of Legacy POPs

Thus far, the influence of anthropogenic climate change on the fate and re-emission of legacy pops has not been well researched or understood. However, there is much evidence to suggest that rising air and water temperatures, ice retreat, and permafrost thawing could promote re-emission of these substances (Ma et al. 2011) and alter their bioaccumulation (Borgå et al. 2010; Ma et al. 2016). Re-emission of POPs such as PCPs and PCDD/Fs from glaciers has already been observed, with the highest concentrations measured in the vicinity of marine-terminating glaciers. Thus, ocean warming has a direct influence (Kobusińska et al. 2020). The gradual decrease or even slight increase in HCB levels in Artic air and ringed seals might also be caused by re-emission (Rigét et al. 2020). Studies from the Chinese Lake Chaohu have shown that the effect of rising temperatures can affect both re-emission and deposition of POPs (Zhang et al. 2019). While the former was observed for DDTs and PAHs, increased biological activity in the lake and the subsequent enhanced sedimentation have led to a reduction in PCB levels (Zhang et al. 2019). In the highest trophic levels of marine food webs, for example, killer whales, a 3% increase in PCB concentration by 2100 was calculated using North Pacific Ocean data, as opposed to a scenario without increased CO_2 emissions (Alava et al. 2017, 2018). Despite these evidence, there are still large gaps in our knowledge on the influence of global transport and release processes on the release of legacy POPs, emphasizing the urgent need for further data collection and monitoring (Nadal et al. 2015; Wang et al. 2019b).

2.4. Outlook

Although the usage of persistent organic pollutants, such as DDTs, PCBs, and PCDDs, were banned by the Stockholm Convention nearly twenty years ago, they can still be detected in decreasing but relevant amounts in marine species, even in pristine areas such as the Arctic and deep-sea trenches. Species such as grey seals and Baltic white eagle, however, whose reproduction rates in the past were drastically reduced by high concentrations of POPs, have been able to recover during the last twenty years.

Causes for concern are a new generation of organohalogen compounds, such as chlorinated paraffines, new halogenated flame retardants and organophosphate esters, which have shown persistence and bioaccumulative potential, and should also be monitored in marine organisms (De Wit et al. 2020). Some of them, such as the short chained chlorinated paraffines, have recently been added to the Stockholm Convention (UNEP 2018a).

Another focus of current research is the interaction between POPs and microplastics. It is considered that microplastics could affect the uptake of POPs by marine organisms (Gassel and Rochmann 2019). Microplastics can accumulate large amounts of hydrophobic organic compounds, and studies have shown that microplastics can be a vector for the transport of POPs into marine organisms (Scopetani et al. 2018). On the other hand, microplastics seem to have relatively low influence on the global POP cycle, as the fraction of POPs sorbed to microplastic is relatively low (Koelmans et al. 2016; Ziccardi et al. 2016). However, some authors have raised the question of whether microplastics themselves should be classified as POPs because of the parallels in their properties, e.g., persistence, long-range transport and accumulation tendencies (Lohmann 2017).

3. Metals as Pollutants in Marine Environments

3.1. Introduction

Compared to artificial pollutants such as plastics, pharmaceuticals, fertilizers or explosives, metals have a special role: as elements, they have always been part of marine cycles and ecosystems. They can neither be created nor eliminated—only released from geological sources or deposited and biologically transformed (UNEP 2018b).

Sources of metal entrance into marine systems are manifold. They are released into the atmosphere by geological processes such as volcanic eruptions or rock weathering; natural cycles have existed long before human impact. These natural cycles are disturbed by the anthropogenic release of metals, e.g., by mining and smelting, combustion of fossil fuels and wastes, or as chemical compounds of various functions, which, in most cases, far exceed the natural release (Nriagu and Pacyna 1988; Dixit et al. 2015). As a result of their persistent nature, removal of metals from the cycles occurs only by sedimentation in the deep seas.

Arsenic is a common element and naturally found in sediments and ocean water. Forty-five kilotons per year (kt/y) are released by natural sources, mainly by volcanic activity and upwelling of deep ocean water. Anthropogenic sources such as mining and smelting of gold and other non-iron metals, and industrial sewage contribute to a comparable amount (Neff 1997). Ocean water has a mean total arsenic concentration of 1.7 µg/L. Arsenic occurs in many inorganic and organic forms with large differences in toxicity and bioaccumulation. Therefore, the total has relatively little informative value (Neff 1997). While inorganic arsenic, which makes the biggest contribution of total arsenic in seawater, is accumulated in low trophic level species, it seems to be transformed to less toxic organic compounds. Most aquatic organisms seem to have a relatively high tolerance for inorganic arsenic (Neff 1997). Edible algae containing large amounts of arsenic could be a threat to human health (Neff 1997). Marine animals seem to accumulate only small amounts of arsenic, as it is excreted rapidly (Neff 2002) and no biomagnification is observed (Rahman et al. 2012). In organisms of higher trophic levels, less arsenic is found mostly as non-toxic arsenobetaine (Rahman et al. 2012).

Due to the toxic effects of cadmium, its uses are relatively limited. Cadmium is only mined as a side product of other metals such as zinc and copper, and often released into the atmosphere. In former times cadmium was used for alloys, electroplating and pigments. However, due to severe nephrotoxic effects, nowadays its use is limited mainly to batteries and semiconductors (Cullen and Maldonado 2013). The annual natural atmospheric release is as low as 0.14–2.5 kt/y, together with anthropological releases from recycling and mining activities of 3 kt/y (WHO 2003). Cadmium shows little to no effect of biomagnification through marine food webs (McGeer et al. 2003). High concentrations are found in low trophic levels such as seaweed, clams and mussels; higher trophic levels contain much less per gram wet weight (w.w.) (Table 1) (Almela et al. 2006; Falcó et al. 2006). In the gland of squids in South-western Atlantic, cadmium levels of up to 1 mg/g w.w. have been found (Lischka et al. 2018). The extreme long half-life of cadmium in the body (>10 years) and its nephrotoxicity make it a threat to humans and marine mammals (Jakimska et al. 2011). Experiments with marine organisms of different trophic levels have shown increasing DNA damage at aquatic concentrations as low as 0.59 µg/L. At higher levels, survival and larval development were found to be reduced (Pavlaki et al. 2016).

Table 1. Arsenic, cadmium, lead and mercury levels in commercially purchased marine species (µg/g wet weight) of different trophic levels. Source: Falcó et al. (2006).

Species	As [µg/g w.w.]	Cd [µg/g w.w.]	Pb [µg/g w.w.]	Hg [µg/g w.w.]
Shrimp	3.85–8.76	0.01–0.03	~0.01	0.02–0.19
Mussel	2.02–2.44	0.02–0.20	0.09–0.21	~0.02
Squid	1.41–4.74	0.05–0.15	~0.01	0.02–0.03
Cuttlefish	2.45–5.33	0.01–0.09	0.01 –0.10	0.04–0.08
Hake	3.22–4.55	<0.01	0.01–0.13	0.12–0.29
Salmon	1.60–2.37	~0.01	0.01–0.25	0.04–0.05
Swordfish	1.78–2.44	~0.01	0.01–0.02	1.59–2.22
Tuna	0.99–1.25	0.01–0.02	0.10–0.02	0.38–0.58

Being one of the most released metallic pollutants of the last century, atmospheric release of lead has dropped rapidly following the ban of tetraethyl-lead as anti-knocking agent in gasoline, in response to its neuro- and reproductive toxicity, and carcinogenic nature (Wani et al. 2015). Like cadmium, it does not biomagnify throughout marine food webs (McGeer et al. 2003). Since the 1970s, efforts have been made to reduce the release of toxic metals, due to awareness of their impact on nature, human health and their economic consequences. The near worldwide ban on leaded gasoline between 1980 and 2017 resulted in a 92% decrease in lead in the surface water of the North Atlantic, down to ~3 ng/L (Boyle et al. 2014; Rusiecka et al. 2018). Between 1990 and 2017, the atmospheric release of lead decreased by 93% (European Environment Agency 2019).

Estimations for the current global mercury emission range from 6500 to 8300 kt/y, including 1900–2900 kt/y from anthropogenic sources, 4600–5300 kt/y from secondary emission of deposited mercury and only 80–600 kt/y from primary geogenic emissions. Artisan, small-scale gold mining and combustion of fossil fuels are the main source of anthropogenic mercury release (UNEP 2018b). Anthropogenic activities might have increased surface ocean mercury levels by 450–660% during the last 600 years and 300% within the last century (Zhang et al. 2014).

Nine-tenths of mercury in the surface water of the ocean originate from atmospheric deposition. Commonly, it is found in the form of inorganic Hg^{II} salts or as elemental Hg^0. Microbiological transformation in sediments and the water column converts inorganic mercury to monomethyl-mercury (MMHg; CH_3Hg^+). MMHg constitutes only a minor fraction of the total aquatic mercury (0.5% in surface water, 1–1.5% in sediments) (Ullrich et al. 2001), but the majority (>95%) of the total mercury in marine organisms. Low aqueous solubility and its lipophilia ensure the accumulation of methyl-mercury in tissues of marine organisms, resulting in a biomagnification progress throughout trophic levels of marine food webs (Mason et al. 1995; McGeer et al. 2003). Lower levels, e.g., algae and mussels, usually contain relatively low concentrations of methyl-mercury, but long-living predatory fish such as swordfish, shark and tuna, or fish consuming mammals such as whales and seals, can accumulate large amounts during their lifetime (Table 2, (FDA 2017). MMHg has strong binding affinities to selenium and sulfur, and forms complexes with cysteine molecules in muscle tissues (Bradley et al. 2017). High concentrations of organic mercury in fish can become a threat to populations, with diets based on the consumption of large quantities of fish. Elevated levels of methyl-mercury can be problematic, as MMHg is neurotoxic (WHO 2003; Myers et al. 2015). The European Food Safety Authority, although highly recommending the consumption of fish, suggests to limit the ingestion of highly loaded fish, e.g., tuna, for children and pregnant women, and advises to choose less contaminated species (EFSA Scientific Committee 2015).

Table 2. FDA monitoring of mercury concentration in commercial fish and shellfish between 1990 and 2012. Source: FDA (2017), no specification whether dry or wet weight.

Species	Mean [µg/g]	Max [µg/g]
Scallop	0.003	0.033
Shrimp	0.009	0.05
Flatfish	0.056	0.218
Herring	0.078	0.56
Snapper	0.17	1.37
Halibut	0.24	1.52
Marlin	0.49	0.92
Bigeye Tuna	0.69	1.82
Shark	0.98	3.22

While the highest amounts of mercury are usually found in long-living predators such as tuna, there are not only differences in concentration based on size and age of individuals, but also between individuals of the same or related species from different ocean basins (de Lacerda et al. 2017; Bezerra et al. 2019). It seems to be correlated with the natural mercury budget and anthropogenic release in the habitat area, as well as with the different production levels of methyl-mercury, based on the availability of organic material for microbial transformation. (Ferriss and Essington 2011; de Lacerda et al. 2017). Comparable observations were made for concentrations of other metals and persistent organic pollutants. Currently, however, the number of studies on this is still relatively small, which makes a more precise assessment difficult (Bezerra et al. 2019).

3.2. Impact of Climate Change on Marine Mercury Release

Another aspect that must be considered when discussing metals as marine pollutants is the impact of climate change on their availability and reemission (Macdonald et al. 2005). Rising temperatures cause melting of glaciers, continental ice and permafrost soils, increasing the availability of organically-bound mercury (de Lacerda et al. 2020). In addition, atmospheric elemental mercury deposits in the soils of the Tundra during snow-free months and is transported by Arctic rivers to the ocean (Obrist et al. 2017). Increased temperature fuels microbial methylation of mercury, resulting in better bioavailability (Emmerton et al. 2013) and higher biomagnification through different trophic levels in the arctic (Schartup et al. 2015; de Lacerda et al. 2020). Comparable observations, but with different mechanisms, have been made in semi-arid coastal areas in Brazil. Reduced annual rainfall and damming of rivers, as well as rising sea levels, extension of saline intrusion and expansion of mangrove areas, result in longer water residence time. This induces extended sulphate reduction mechanisms and elevated production of dissolved organic carbon, which is able to form complexes with mercury, making mercury bioavailable. As a result, ten times elevated mercury concentrations were found in shrimps *L. vannamei* from the estuary compared to individuals from upstream regions (de Lacerda et al. 2020). Increasing levels of mercury have also been observed, starting from the mid-1990s in tuna caught in the North Pacific Ocean. Higher temperatures could alter metabolic processes, thus resulting in elevated mercury uptake into biota as well as changes of marine food webs (Grieb et al. 2020).

3.3. Metal Pollution in the Baltic Sea

Metal pollution is one of the issues being monitored by the Baltic Marine Environment Protection Commission (HELCOM). In sediment core samples from the Bothnian Bay, reflecting the heavy metal levels of almost a century, decreasing levels of Cd, Hg, and Pb have been found for the last 40 years. In contrast, arsenic remains at an elevated level of 50 mg/kg, with even an uncertain spike of >100 mg/kg in the mid-1990s. Levels have declined, but remain at 50 mg/kg (Vallius 2014). In herring livers and blue mussel soft bodies, decreasing lead concentrations were observed between 1998 and 2015 (HELCOM 2018). The highest concentrations in 2015 were found in the Gulf of Finland (0.203 µg/g w.w. herring liver) and lowest concentrations in the Kattegat (0.01 µg/g w.w.). Even with a decreased atmospheric release of some 73% less mercury in the European Union between 1990 and 2017 (European Environment Agency 2019), only 18 out of 66 biota sample data sets (herring muscle) indicated falling levels of MMHg. The majority still exceeded the threshold value (20 µg/g w.w.), and in five places values had even risen (HELCOM 2018). For cadmium, the annual release in the European Union had fallen to 35% between 1990 and 2017 (European Environment Agency 2019). However, while Cd in the tissue of blue mussels from the Gdansk Basin between 2000 and 2016 had dropped by 60%, down to ~0.15 µg/g w.w., mussels from the Kattegat had accumulated 25% higher amounts between 1997 and 2013 (HELCOM 2018).

3.4. Munitions as a Source of Mercury and Arsenic in the Baltic Sea

Concerning the Baltic Sea, another repository of metallic pollution must be mentioned. In the aftermath of the First and Second World Wars, quick solutions had to be found to get rid of huge stockpiles of both conventional and chemical weapons, and munitions. This problem affected all participating parties, but was focused on disarming the defeated Germans. Due to the enormous amounts, it was decided that the easiest and safest solution was to just dump the warfare material into the North and Baltic Seas (Bełdowski et al. 2020). Ecological concerns were much smaller than the fear of misuse. Many of these weapons contained not only conventional explosives, but also compounds such as mercury fulminate ($Hg(CNO)_2$) (Bełdowski et al. 2019) in detonators and various chloro-organic chemical warfare agents such as Lewisite ($C_2H_2AsCl_3$) or Clark I and II (I: [$(C_6H_5)_2AsCl$], II: [$(C_6H_5)_2AsCN$]) (HELCOM 1994; Garnaga et al. 2006).

At the end of the twentieth century, the fear arose that these highly toxic arsenic compounds could leak from ammunitions and accumulate in biota and sediments. Even degradation would still lead to toxic, inorganic compounds (HELCOM 1994). Different studies on sediments close to munition dumping sites indicated that this process might have already happened to a small extent. Further studies are urgently needed and monitoring systems are to be developed to understand and follow the imminent consequences of this problem (Garnaga et al. 2006; Bełdowski et al. 2016).

For conventional munitions, Bełdowski et al. determined that 0.1% of the approximated >300.000 tons dumped to the Baltic Sea could be mercury, which is more than 300 tons in total. A conclusion from their findings was that, while the amount of mercury leaking from conventional weapons is relatively hard to evaluate from other sources, it may become a measurable source when land-based release further decreases (Bełdowski et al. 2019) (see Section 4 for further details on munitions in the seas).

3.5. Outlook

The public awareness of metals as pollutants and their ecological and human health impact has increased during the last 50 years. National and international contracts have shrunk the anthropogenic emission of arsenic, cadmium, mercury and lead in Europe and North America, although emission remained high and even increased in Latin America, Africa and Asia. The near world-wide ban of leaded fuel led to significantly decreasing amounts of atmospheric lead. The release of mercury, however, is still high as a result of small-scale gold mining and combustion of fossil fuels. Due to the slow elimination of metals by sedimentation from the global system, it will take decades to reach a pre-industrial level, even with no new emission. The biomagnification of monomethyl-mercury, with high levels found in long-living predatory fish and mammals, remains problematic. To face the worldwide mercury release problem, in August 2017, the United Nations Minamata convention came to power; aiming on achieving lower mercury emissions by banning mercury goods, and working on solutions to decrease the mercury output of gold mining and combustion (UNEP 2019). Climate change is an additional driver of marine mercury release, as rising temperatures favor the re-emission of mercury. This can already be observed in the Arctic as well as in semi-arid areas of the subtropics. Furthermore, dumped munitions contribute as a source of mercury and arsenic in marine environments.

4. Munitions in Seas

4.1. Introduction

The intensive exploitation of the oceans by humans, overfishing and the discharge of hazardous substances pose great risks to marine ecosystems. Unfortunately, and in addition, the seas worldwide are threatened by a relatively new source of pollution. Millions of tons of all kinds of munitions are dumped into seas worldwide during and after war actions, e.g., First and Second World Wars. Besides the risk of detonation with increased human access (fisheries, cable constructions, wind farms and pipelines, ship traffic, tourism), leaching and distribution of toxic chemicals from corrosive munitions may accumulate in marine organisms, enter the marine food chain and directly affect human health. The usage of munitions increased especially at the beginning of the twentieth century and then had its summit during the World Wars. The reason the importance of dumped munitions nowadays, i.e., decades after its intentional disposal, is bigger than ever is the fact that metal munition housings are starting to corrode, thereby releasing their contaminants into the environment. Munitions contain different groups of hazardous substances such as organic explosive compounds, chemical warfare agents, various types of metals and other munition-structural components. Unfortunately, little is known about the fate of these components in the marine environment, and their behavior is only poorly understood regarding possible health effects on humans; knowledge on seafood, such as various kinds of fish, mussels and crustaceans that are consumed worldwide and may contain conventional explosives or chemical warfare agents is rather low. Howsoever, because of the extensive corrosion of the metal shells, the release of toxic compounds into the water column will increase over time (Beddington and Kinloch 2005; Juhasz and Naidu 2007; Beck et al. 2018).

It is proven that explosive chemicals such as 2,4,6-Trinitrotoluene (TNT) and its derivatives are known for their toxicity and carcinogenicity (Bolt et al. 2006). Today, already measurable readings of explosive residues are detected in biota from the vicinity of the dumped munitions such as sea mines and others, a fact that already indicates the entry of these compounds into the marine food chain. Organic chemicals, as components of munitions, are transformed by a variety of mechanisms, abiotic and biotic, to products which are often no less toxic than their parent compounds. According to our current knowledge, complete degradation to harmless compounds, based on their chemical structures, e.g., nitroaromatics and nitramines, is not expected. Rather, they persist in the environment, for example, by sorption on sediments and other particles, and accumulate in plants and various types of animals as well as in microalgae and bacteria (Rosen and Lotufo 2010; Beck et al. 2018).

4.2. Explosives and Chemical Warfare Agents in the Marine Environment

In recent years, munitions in the seas have increasingly become the focus of research, and several studies have proven that munition compounds such as explosives and chemical warfare agents enter the marine environment. One important study was published in 2004 by Porter, Barton, and Torres about the naval gunnery and bombing range on the eastern end of Isla de Vieques, Puerto Rico, which took place between 1943 and 2003. Here, it was proven that the coral reefs are littered with leaking unexploded ordnance (UXO). In this study, samples were taken in the vicinity of an unexploded 2000-pound air-dropped bomb which was about to corrode. The data unequivocally show that toxic substances leaching from unexploded ordnance have entered the coral reef marine food web. TNT and other explosive chemicals have been measured in water, sediment and biota, for example, in dusky damselfish (*Stegastes adustus*), sea urchin (*Diadema antillarum*) and different kinds of corals (Porter et al. 2011).

However, reports of explosives found in marine biota are not limited to the North Atlantic Ocean. In particular, the North Sea and Baltic Sea both belong to the largest areas contaminated with munitions. For instance, the German parts of the Baltic Sea and North Sea alone contain an estimated amount of 1.6 million metric tons of munitions (Böttcher et al. 2011). Gledhill et al. (2019) found several types of explosive chemicals in marine biota such as algae, asteroidea and tunicata which had been collected at Kolberger Heide, a known dumping ground for different types of munitions in the Kiel Bight in the Baltic Sea. The Kolberger Heide is a section of the western Kiel Bight at the entrance to Kiel Fjord, Germany, with a size of approximately 1.260 ha, located at a distance of three to five nautical miles to the shoreline. It was used as an area for dumping munitions after World War II. In addition, in the last years, some torpedo heads and mines have been destroyed selectively in Kolberger Heide by blasting (Böttcher et al. 2011). Gledhill and her team found body burdens of octahydro-1,3,5,7-tetranitro-1,3,5,7-tetreazocine (HMX), hexahydro-1,3,5-trinitro-1,3,5-triazine (RDX), 2,4,6-trinitrotoluene (TNT) and ten other explosives with concentrations up to the highest of nearly 25 $\mu g/g$ in starfish.

In two case studies, also carried out in Kolberger Heide, blue mussels (*Mytilus* spp.) were deployed selectively at corroding moored mines or loose hexanite lying on the seafloor. After three months, in the mussels deployed at the moored mines, body burdens of 4-aminodinitrotoluene (4-ADNT), a degradation product of TNT, was found—up to 10 ng/g mussel tissue (wet weight) (Appel et al. 2018). In mussels directly deployed at lumps of loose hexanite, 4-ADNT, 2-aminodinitrotoluene (2-ADNT) and TNT itself were found at total concentrations up to 260 ng/g mussel tissue (wet weight) (Strehse et al. 2017). Blue mussels are one of the most common seafood species worldwide and with these studies it has been proven that explosive chemicals may enter the marine and human food chain even after a short exposure period.

4.3. Munition-Related Chemicals in Seafood

Even though the presence of all kinds of munitions in seas has been known for decades, only few reports exist about explosives that have been determined in marine seafood species until now, especially compared to other sources of pollutants. The situation looks even worse with chemical warfare agents, because of the limited data. Chemical warfare agents were produced on a large-scale during World Wars I and II, and have also been disposed in the seas. In the Baltic Sea, for example, in the Bornholm Deep and the Gotland Basin as well as in the Skagerrak between Denmark and Norway, large dumping operations with chemical weapons took place. Within the Baltic Sea area 40,000 tons and in the Skagerrak 168,000 tons have been dumped, mainly arsenic-containing agents and sulfur mustard. Niemikoski et al. (2017) have published the first study that reports the occurrence of oxidation products of Clark I and/or II in marine biota. Clark I and II belong to a group of high toxic compounds, which were used during the trench warfare of World War I. Niemikoski and her team found the degradation products in lobster (*Nephrops norvegicus*) and a flatfish species collected at Måseskär dumpsite, an area in the eastern Skagerrak (Niemikoski et al. 2017). However, only trace concentrations below the limit of quantification were detected, but laboratory exposure studies with chemical warfare agents demonstrated that indeed those compounds enter the marine biota, as has been demonstrated by Höher et al. (2019).

4.4. Toxicological Aspects

Conventional munitions and chemical warfare agents are problematic from the toxicological point of view, for humans and the ecosphere. Besides the direct acute toxic effects of chemical warfare agents on humans (which were intended in war actions), chronic effects are yet to be considered as well for all groups of chemicals used in munitions. The expected effects could be highly diverse and there are fears that marine microorganisms as well as mammals or other higher organisms will be severely affected. In general, marine biota may be negatively influenced in their behavior, health, fitness, growth and germination. Some organisms, for example, mussels, seem to be more robust regarding chemicals used in munitions than other invertebrates as well as vertebrates such as fish (Rosen and Lotufo 2007). Additionally, differences must be considered between effects on adult or juvenile animals. For example, acute toxicity effects on shrimps have been observed at 0.98 mg/L TNT in water. On mollusks, sub-lethal effects such as embryo development have been observed at 0.75 mg/L TNT (Beck et al. 2018). Such free water concentrations have been found near unexploded ammunitions in the areas of Isla de Vieques, Puerto Rico and at Kolberger Heide, Germany (Porter et al. 2011; Beck et al. 2019).

Whereas the accumulation of TNT and RDX has been reported for marine flora and fauna species, it has also been suggested that different species of the marine biota metabolize munition compounds. There is some evidence that metabolites and transformation products, e.g., the TNT metabolites 2-ADNT and 4-ADNT, may be more toxic than the parent compound TNT. Furthermore, explosives and chemical warfare agents have recently been suspected to have sub-lethal genetic effects on marine organisms at relatively low water concentrations. It has been shown that significant changes in gene transcript expression took place in freshwater minnows after exposure to the explosive chemical RDX (Gust et al. 2011). Elevated cytotoxicity and genotoxicity have been observed in fish collected in the Baltic Sea, near munitions dumpsites (Beck et al. 2018).

4.5. Latest Research Activities

In recent years, research activities on the fate and effects of dumped munitions have increased considerably. At first, the main focus was on chemical warfare agents. In 2007 the "CHEMSEA" project started with the aim to learn more about the locations of dumping areas in the Baltic Sea, the content and state of the chemical munitions, and how these respond to Baltic conditions.[1] The project was completed in 2013 and directly followed by the project "MODUM" (Towards the Monitoring of Dumped Munitions Threat). The goal of this project was the establishment of a monitoring network observing chemical weapon dumpsites in the Baltic Sea.[2] The projects "DAIMON" and "DAIMON 2" (Decision Aid for Marine Munitions), deal with the question on how to proceed with mapped and identified warfare objects. It will develop tools to support the governments in the Baltic Sea Region to help with decision-making on whether remediation is needed for both dumped conventional munitions and chemical warfare agents. "DAIMON 2" will end up in early 2021.[3]

Between 2016 and 2019, "UDEMM" (Environmental Monitoring for the Delaboration of Munitions in the Sea) performed the first scientific monitoring in a known dumping area for conventional munitions.[4] The aim of this multidisciplinary approach was to identify the precise location of UXOs (unexploded ordnances) by detection with high-resolution methods such as multi beam, oceanographic mapping and modeling, analyzing sediment, water and biota for explosive chemicals, and to establish a biomonitoring system with blue mussels to observe a possible contamination of the environment with explosives. Importantly, the UDEMM project was closely connected to the project RoBEMM (Robotic Underwater Salvage and Disposal Process) with the technology to remove explosive ordnance in the sea, in particular in coastal and shallow waters. In 2018 the "North Sea Wrecks" (NSW) project started its work and is currently dealing with wrecks and munitions in the North Sea. NSW will provide tools for planners, economic actors and other stakeholders to assess and propose solutions for risk mitigation.[5]

4.6. Outlook

In conclusion, it is meanwhile conceivable that munitions in the sea are a worldwide problem which has been ignored by the society for decades. Since the increased use of the oceans, munitions in seas have increasingly become a problem for the commercial society, such as the energy and communication sectors, because of disturbances during the building of wind farms, pipelines and cable constructions.

---

[1]　See https://ec.europa.eu/regional_policy/en/projects/germany/chemsea-tackles-problem-of-chemical-munitions-in-the-baltic-sea (accessed on 19 January 2021).

[2]　See www.nato.int/cps/en/natohq/news_136380.htm (accessed on 19 January 2021).

[3]　See www.daimonproject.com (accessed on 19 January 2021).

[4]　See https://udemm.geomar.de/ (accessed on 19 January 2021).

[5]　See www.dsm.museum/forschung/forschungsprojekte/north-sea-wrecks (accessed on 19 January 2021).

In recent years, it also became more and more visible that dumped munitions have started to corrode, and chemical constituents of these munitions now leak into the environment and affect flora and fauna. Studies on possible human uptake of these chemicals from contaminated fish or seafood are still pending. There are also initial indications that increasing global warming is accelerating the progress of corrosion of the metal casings of the munitions, and that the chemicals contained in the munitions are increasingly dissolving in water. Fortunately, society is starting to face this problem and has increased research activities with focus on munitions in the sea in the last few years. Nevertheless, until now, too little is known about the occurrence, fate, and effects of munition-related chemical contaminants and their impact on the environment and human health. To prevent an increasing ecological risk, the need for additional research, the closing of knowledge gaps and a better understanding of threatening long-term effects on the ecosphere and human health are mandatory.

5. Pharmaceuticals in the Marine Environment

5.1. Introduction

Over the past centuries, alleviation and cure of diseases have remained an important part of human activities. In addition to the use of plant, animal and mineral ingredients or even human raw materials for centuries, synthetic active substances have become an integral part of modern medical therapy since the middle of the 19th century. Active compounds used in medical regimens for human and animals often have powerful effects and benefits in many areas of application. It is estimated that more than 2000 pharmaceutically active ingredients are currently in use worldwide (Bergmann et al. 2011). Unfortunately, approximately half of them have an impact on the environment. The chemical composition of pharmacologically active compounds is wide-ranging, from small, fairly simple to complex molecular structures, even some of them of a natural origin. Within the last few years, another group of therapeutics—the so-called biologicals—are entering the pharmaceutical markets. Biologicals are characterized by their macromolecular structure such as protein- and nucleic acid-based compositions. Medicines are used in humans and animals, as well as in fruit growing. Pharmaceuticals comprise a wide range of different therapeutic drug classes such as pain killers, antibiotics, cardiovascular affecting drugs, antidepressants, endocrinologically active drugs, and many more. Each group consists of a multitude of active ingredients, sometimes with completely different chemical properties (Kümmerer 2004).

5.2. Pathways into the Environment

There are many ways of entries into the environment. In European countries and the United States, emissions during manufacturing, transport and storage are probably low because of strict regulations in Good Manufacturing Practices (GMPs) (Kümmerer 2004). The situation looks different in developing countries, depending on the manufacturer and/or national regulations. The majority of active ingredients enter the environment through their application and upon treatment of human, animal or plant diseases (Gaw et al. 2014). After ingestion, the drugs are more or less metabolized in the body, depending on their chemical composition. For example, the painkiller acetylsalicylic acid is largely metabolized, while iodinated X-ray contrast agents such as iopamidol are relatively inert and, therefore, persistent in the environment. Nevertheless, metabolites of active compounds may not only have unwanted side effects within the target organism but may also affect the environment as well. Excretion of medical drugs from their target organisms occurs via feces and/or urine, where they are flushed through toilets into the sewage and finally introduced into sewage treatment plants. Alarmingly, and in addition, lots of unused liquid pharmaceutical medicines in households are disposed down the drain, even in industrial nations. It is estimated that nearly one-third of the total volume of all pharmaceuticals sold in Germany is disposed via household waste or down the drains (Greiner and Rönnefahrt 2003). In the optimal case, the wastewater is cleaned in sewage treatment plants before entering rivers or the sea.

Unfortunately, many active ingredients such as the nonsteroidal anti-inflammatory drug diclofenac are only poorly broken down in sewage treatment plants, and thus enter the subsequent water compartments more or less undegraded. Especially when applying drugs in commercial animal husbandry and aquaculture, leftovers enter aquatic systems by washouts from fields because of the use of manure and sewage sludge to fertilize fields. In aquaculture systems, the contamination occurs even directly because fish farms are not necessarily separated from the surrounding natural environment (Gräslund et al. 2003; de Lacerda et al. 2019). The entry into the environment also depends on country and regional specific differences, e.g., whether sewage is treated by sewage treatment plants, household garbage is disposed only at landfills or burned in waste incineration plants. There are also differences in the worldwide consumption of active pharmaceutical compounds, for instance, whether antibiotics are only available on prescription as well as the preference for specific therapeutic options, such as the use of contraceptive pills or strong painkillers such as opioids (Kümmerer 2004).

In terms of the huge variety of active compounds and pharmaceutical drugs, the impacts on the marine environment are neither estimated so far nor predictable (Kümmerer 2004). Every medication can be considered in terms of environmental sustainability from different points of view. Regional differences in the amount of active compounds being used are of high interest as well as the specific characteristics such as the toxicity of a compound within the environment or its persistence. Almost impossible to assess is the "cocktail effect" of a mixture of different active pharmaceutical ingredients even in small amounts (Vasquez et al. 2014). In medical therapy for a single person or animal, only a few compounds are used at the same time, such that intended effects and unwanted side effects are comparably well predictable. Nevertheless, the same components released into the environment, whether by the treated person or animal itself, may affect various types of organisms of different trophic levels, a fact which, in certain circumstances, can have severe impacts on the flora and fauna of the ecosphere (Álvarez-Muñoz et al. 2015).

5.3. Occurrence and Effects of Active Ingredients in the Environment—Examples

In the 1970s, the negative effects of pharmaceuticals on the environment started to become a focus of public awareness and even so of scientific interest. First attention was given to the hormones. Since the mid-1990s the occurrence, fate and effects of active hormonal compounds used in pharmaceuticals were recognized, such that more and more scientific activities had been initiated, especially in the United States and Europe. One of the first findings was the conclusion that hormones are not easily biodegraded. From then on, it was only a matter of time until several studies proved adverse effects of these compounds on the marine environment (Kümmerer 2004). For example, nowadays the feminization of male fish resulting from exposure to estrogens is well known (Sumpter 1995; Zeilinger et al. 2009). In addition, the connection between the undesirable increase in microbial resistances and the high and uncritical application of antibiotics in human and veterinary medicine is irrefutable (Cabello 2006). Especially, overly short treatment regimens and/or non-therapeutic dosages released into the environment, e.g., contained in feces or manure of treated humans or animals, respectively, help to develop bacterial resistance mechanisms.

In the meantime, a large number of worldwide studies on the presence of pharmaceutically-active compounds in the marine environment are available. The occurrence of active ingredients is not only described in compartments such as water and sediment, but also in aquatic plants and various animal species such as fish and clams (Álvarez-Muñoz et al. 2015). For example, a study published in 2015 has shown the extreme wide spread of pharmaceutical compounds in different marine species from coastal areas in Europe, and the high rate of occurrence and amount of different compounds found in species collected from the same area (Álvarez-Muñoz et al. 2015). Fish, bivalves and macroalgae, which are in addition also common seafood species, have been analyzed for possible contaminations with one or more out of 35 pharmaceutical compounds from different therapeutic groups, such as antibiotics, anti-inflammatories or psychiatric drugs. The species have been collected from potentially contaminated areas in Italy, the Netherlands, Spain, Portugal and Norway. It has been shown that 16 different active ingredients were found in bivalves collected in Portugal, Italy and Spain. The analysis of fish collected in Portugal and the Netherlands revealed the occurrence of ten active compounds (Álvarez-Muñoz et al. 2015). To the authors, knowledge of this was the first time that pharmaceutically-active compounds had been detected in marine fish. In macroalgae collected from Fureholmen, Solund (Norway), four active ingredients have been found (Álvarez-Muñoz et al. 2015). This is particularly remarkable because there is little to no human habitation or industry nearby the location, but nevertheless, the betablockers metoprolol and propranolol have been detected as well as the antibiotic azithromycin, and the psychiatric drug diazepam. The most recurring substances were the psychiatric drug venlafaxine, the diuretic hydrochlorothiazide, the betablocker metoprolol and the antibiotic azithromycin. The highest levels were measured for venlafaxine and azithromycin with a body burden up to 36.0 ng/g dry weight and up to 13.3 ng/g dry weight, respectively. This study is only one example of a large number of studies which have shown the wide distribution of pharmaceutically-active compounds in the marine environment (Álvarez-Muñoz et al. 2015).

5.4. Prevention of Entries

Due to the fact that most of the active ingredients enter the marine environment as a consequence of the therapeutic use of drugs, is it impossible that future entries can be completely prevented. Stricter rules in use and prescription of drugs could help to mitigate the problem, especially with a view on the often too uncritical application of antibiotics in the human and veterinarian use (Hulscher et al. 2010). In the future an environmentally friendly disposal should also be considered, particularly with regard to the disposal of unused medicines in private households and hospitals, or the use of manure as fertilizer in agriculture. The most important role could be played by sewage treatment plants. The construction of wastewater treatment plants worldwide and the improvement of already existing technologies could drastically reduce the input of these substances into the marine environment.

Obviously and unfortunately, pharmaceutically-active compounds are now being distributed worldwide in rivers and seas, but the risks to the environment due to their complexity cannot be fully ascertained. Therefore, it is advisable to carry out monitoring programs. For example, in 2013 the European Union included the nonsteroidal anti-inflammatory drug diclofenac, and the synthetic hormones ethinylestradiol and β-estradiol in the European monitoring list (Directive 2013/39 2013). In 2017 the Baltic Marine Environment Protection Commission (HELCOM) also decided to choose diclofenac and estrogen to be used as HELCOM indicators of the health of the Baltic Sea ecosystem. It is intended to meet the requirements of the Marine Strategy Framework Directive (2008/56/EC). These examples show the efforts to better understand the fate and distribution of pharmaceutically-active compounds in the seas.

Despite their intended beneficial therapeutic effects, it has to be feared that a high number of active ingredients in pharmaceuticals could impact the marine environment. Since the effects on the ecosystems cannot be fully estimated, the entry of pharmaceutical compounds into the marine environment, rivers and lakes as well as in groundwater and drinking water should be avoided or at least limited, even if the compounds in question have low acute toxicity in humans.

6. General Conclusions

The examples described, of course, represent only a part of the pollutants occurring in the marine environment. This chapter could otherwise be continued indefinitely. Endocrine disruptors and petroleum derivatives also play an important role in water pollution. Some of them, such as POPs, microplastics and metals, have long been known to the general public and politics. Others, such as munitions in the seas and pharmaceuticals, have not yet come into full focus.

Assessing the risks posed to humans and the environment by such pollutant sources is, in particular, a major challenge. On one hand, substances can be measurable even decades after their ban, and also in regions far from the source of contamination. On the other hand, the "cocktail effect" plays an important role in the assessment of toxic effects, especially on marine flora and fauna, which should not be underestimated. Transport processes such as the adhesion of pollutants to microplastics and their possible effects should be considered as well.

Reducing the input of pollutants into the environment should therefore be a highly recognized goal worldwide and for the benefit of all, both for human health and the maintenance of intact ecosystems.

Author Contributions: J.S.S. wrote sections "Munitions in Seas" and "Pharmaceuticals in the Marine Environment, T.H.B. wrote sections "Persistent Organic Pollutants (POPs)" and "Metals as Pollutants in the Marine Environment", E.M. wrote section "Microplastics in the Marine Environment". All authors have read and agreed to the published version of the manuscript.

Funding: This research received no external funding.

Conflicts of Interest: The authors declare no conflict of interest.

References

Airaksinen, Riikka, Anja Hallikainen, Panu V. Rantakokko, Päivi H. Ruokojärvi, Pekka J. Vuorinen, Raimo Parmanne, Matti Verta, Jaakko Mannio, and Hannu A. Kiviranta. 2014. Time trends and congener profiles of PCDD/Fs, PCBs, and PBDEs in Baltic herring of the coast of Finland during 1978–2009. *Chemosphere* 114: 165–71. [CrossRef]

Akindele, Emmanuel O., and Chibuisi G. Alimba. 2021. Plastic pollution threat in Africa: Current status and implications for aquatic ecosystem health. *Environmental Science and Pollution Research Technology* 28: 7636–51. [CrossRef]

Alava, Juan José, William W. L. Cheung, Peter S. Ross, and U. Rashid Sumaila. 2017. Climate Change–Contaminant Interactions in Marine Food Webs: Toward a Conceptual Framework. *Global Change Biology* 23: 3984–4001. [CrossRef] [PubMed]

Alava, Juan José, Andrés M. Cisneros-Montemayor, U. Rashid Sumaila, and William W. L. Cheung. 2018. Projected Amplification of Food Web Bioaccumulation of MeHg and PCBs under Climate Change in the Northeastern Pacific. *Scientific Reports* 8: 13460. [CrossRef]

Almela, Concepción, M Jesús Clemente, Dinoraz Vélez, and Rosa Montoro. 2006. Total Arsenic, Inorganic Arsenic, Lead and Cadmium Contents in Edible Seaweed Sold in Spain. *Food and Chemical Toxicology* 44: 1901–08. [CrossRef] [PubMed]

Álvarez-Muñoz, Diana, Sara Rodríguez-Mozaz, Ana Luísa Maulvault, Alice Tediosi, Margarita Fernández-Tejedor, Fredericus Van den Heuvel, Michiel Kotterman, Aantónio Marques, and Damià Barceló. 2015. Occurrence of Pharmaceuticals and Endocrine Disrupting Compounds in Macroalgaes, Bivalves, and Fish from Coastal Areas in Europe. *Environmental Research* 143: 56–64. [CrossRef]

Andrades, Ryan, Roberta Aguiar dos Santos, Agnaldo Silva Martins, Davi Teles, and Robson Guimarães Santos. 2019. Scavenging as a pathway for plastic ingestion by marine animals. *Environmental Pollution* 248: 159–65. [CrossRef]

Appel, Daniel, Jennifer S. Strehse, Hans-Jörg Martin, and Edmund Maser. 2018. Bioaccumulation of 2,4,6-Trinitrotoluene (TNT) and Its Metabolites Leaking from Corroded Munition in Transplanted Blue Mussels (M. Edulis). *Marine Pollution Bulletin* 135: 1072–78. [CrossRef] [PubMed]

Ashton, Karen, Luke Holmes, and Andrew Turner. 2010. Association of Metals with Plastic Production Pellets in the Marine Environment. *Marine Pollution Bulletin* 60: 2050–55. [CrossRef]

Barboza, Luís Gabriel Antão, A. Dick Vethaak, Beatriz R. B. O. Lavorante, Anne-Katrine Lundebye, and Lúcia Guilhermino. 2018. Marine Microplastic Debris: An Emerging Issue for Food Security, Food Safety and Human Health. *Marine Pollution Bulletin* 133: 336–48. [CrossRef] [PubMed]

Beck, Aaron J., Martha Gledhill, Christian Schlosser, Beate Stamer, Claus Böttcher, Jens Sternheim, Jens Greinert, and Eric P. Achterberg. 2018. Spread, Behavior, and Ecosystem Consequences of Conventional Munitions Compounds in Coastal Marine Waters. *Frontiers in Marine Science* 5: 141. [CrossRef]

Beck, Aaron J., Eefke M. van der Lee, Anja Eggert, Beate Stamer, Martha Gledhill, Christian Schlosser, and Eric P. Achterberg. 2019. In Situ Measurements of Explosive Compound Dissolution Fluxes from Exposed Munition Material in the Baltic Sea. *Environmental Science & Technology* 53: 5652–60.

Beddington, John, and Anthony J. Kinloch. 2005. *Munitions Dumped at Sea: A Literature Review*. London: Imperial College London Consultants, Available online: http://www.environet. eu/pub/pubwis/rura/000ic_munitions_seabed_rep.pdf (accessed on 19 January 2021).

Bełdowski, Jacek, Marta Szubska, Emelyan Emelyanov, Galina Garnaga, Anna Drzewińska, Magdalena Bełdowska, Paula Vanninen, Anders Östin, and Jacek Fabisiak. 2016. Arsenic Concentrations in Baltic Sea Sediments Close to Chemical Munitions Dumpsites. *Deep Sea Research Part II: Topical Studies in Oceanography* 128: 114–22. [CrossRef]

Bełdowski, Jacek, Marta Szubska, Grzegorz Siedlewicz, Ewa Korejwo, Miłosz Grabowski, Magdalena Bełdowska, Urszula Kwasigroch, Jacek Fabisiak, Edyta Lońska, Mateusz Szalad, and et al. 2019. Sea-Dumped Ammunition as a Possible Source of Mercury to the Baltic Sea Sediments. *Science of The Total Environment* 674: 363–73. [CrossRef]

Bełdowski, Jacek, Matthias Brenner, and Kari K. Lehtonen. 2020. Contaminated by war: A brief history of sea-dumping of munitions. *Marine Environmental Research* 162: 105189. [CrossRef]

Bergmann, Axel, Reinhard Fohrmann, and Frank-Andreas Weber. 2011. *Zusammenstellung von Monitoringdaten zu Umweltkonzentrationen von Arzneimitteln*. UBA TEXTE 66/2011. Dessau-Roßlau: UBA, ISSN 1862-4804.

Bertazzi, Pier Alberto, and Alessandro di Domenico. 1994. Chemical, Environmental, and Health Aspects of the Seveso, Italy, Accident. *Dioxins and Health* 1: 587–632.

Besseling, Ellen, Paula Redondo-Hasselerharm, Edwin M. Foekema, and Albert A. Koelmans. 2019. Quantifying ecological risks of aquatic micro- and nanoplastic. *Critical Reviews in Environmental Science and Technology* 49: 32–80. [CrossRef]

Bezerra, Moises F., Luiz D. Lacerd, and Chun-Ta Lai. 2019. Trace metals and persistent organic pollutants contamination in batoids (Chondrichthyes: Batoidea): A systematic review. *Environmental Pollution* 248: 684–95. [CrossRef] [PubMed]

Boerger, Christiana M., Gwendolyn L. Lattin, Shelly L. Moore, and Charles J. Moore. 2010. Plastic Ingestion by Planktivorous Fishes in the North Pacific Central Gyre. *Marine Pollution Bulletin* 60: 2275–78. [CrossRef]

Bolt, Hermann M., Gisela H. Degen, Susanne B. Dorn, Sabine Plöttner, and Volker Harth. 2006. Genotoxicity and Potential Carcinogenicity of 2,4,6-TNT Trinitrotoluene: Structural and Toxicological Considerations. *Reviews on Environmental Health* 21: 217–228. [CrossRef] [PubMed]

Bordós, Gábor, Szilveszter Gergely, Judit Háhn, Zoltán Palotai, Éva Szabó, Gabriella Besenyő, András Salgó, Péter Harkai, Balázs Kriszt, and Sándor Szoboszlay. 2021. Validation of pressurized fractionated filtration microplastic sampling in controlled test environment. *Water Research* 189: 116572. [CrossRef] [PubMed]

Borgå, Katrine, Tuomo M. Saloranta, and Anders Ruus. 2010. Simulating Climate Change-Induced Alterations in Bioaccumulation of Organic Contaminants in an Arctic Marine Food Web. *Environmental Toxicology and Chemistry* 29: 1349–57. [CrossRef]

Böttcher, Claus, Tobias Knobloch, Niels-Peter Rühl, Jens Sternheim, Uwe Wichert, and Joachim Wöhler. 2011. *Munitionsbelastung Der Deutschen Meeresgewässer-Bestandsaufnahme Und Empfehlungen (Stand 2011)*. Meeresumwelt Aktuell Nord- und Ostsee, 2011/3. Hamburg and Rostock: Bundesamt für Seeschifffahrt und Hydrographie (BSH).

Bowley, Jake, Craig Baker-Austin, Adam Porter, Rachel Hartnell, and Ceri Lewis. 2021. Oceanic Hitchhikers—Assessing Pathogen Risks from Marine Microplastic. *Trend in Microbiology* 29: 107–16. [CrossRef]

Boyle, Edward A., Jong-Mi Lee, Yolanda Echegoyen, Abigail Noble, Simone Moos, Gonzalo Carrasco, Ning Zhao, Richard Kayser, Jing Zhang, Toshitaka Gamo, and et al. 2014. Anthropogenic Lead Emissions in the Ocean: The evolving global experiment. *Oceanography* 27: 69–75. [CrossRef]

Bradley, Mark, Benjamin Barst, and Niladri Basu. 2017. A Review of Mercury Bioavailability in Humans and Fish. *International Journal of Environmental Research and Public Health* 14: 169. [CrossRef]

Browne, Mark A., Awantha Dissanayake, Tamara S. Galloway, David M. Lowe, and Richard C. Thompson. 2008. Ingested Microscopic Plastic Translocates to the Circulatory System of the Mussel, *Mytilus edulis* (L.). *Environmental Science & Technology* 42: 5026–31.

Bucci, Kennedy, Matthew Tulio, and Chelsea. M. Rochman. 2020. What is known and unknown about the effects of plastic pollution: A meta-analysis and systematic review. *Ecological Applications* 30: e02044. [CrossRef] [PubMed]

Cabello, Felipe C. 2006. Heavy Use of Prophylactic Antibiotics in Aquaculture: A Growing Problem for Human and Animal Health and for the Environment. *Environmental Microbiology* 8: 1137–44. [CrossRef]

Cappello, Tiziana, Giuseppe De Marco, Gea Oliveri Conti, Alessia Giannetto, Margherita Ferrante, Angela Mauceri, and Maria Maisano. 2021. Time-dependent metabolic disorders induced by short-term exposure to polystyrene microplastics in the Mediterranean mussel Mytilus galloprovincialis. *Ecotoxicology and Environmental Safety* 209: 111780. [CrossRef] [PubMed]

Chagnon, Catherine, Martin Thiel, Joana Antunes, Joana Lia Ferreira, Paula Sobral, and Nicolas Christian Ory. 2018. Plastic ingestion and trophic transfer between Easter Island flying fish (*Cheilopogon rapanouiensis*) and yellowfin tuna (*Thunnus albacares*) from Rapa Nui (Easter Island). *Environmental Pollution* 243, Part A: 127–33. [CrossRef]

Claessens, Michiel, Steven De Meester, Lieve Van Landuyt, Karen De Clerck, and Colin R. Janssen. 2011. Occurrence and Distribution of Microplastics in Marine Sediments along the Belgian Coast. *Marine Pollution Bulletin* 62: 2199–204. [CrossRef] [PubMed]

Cole, Matthew, Pennie Lindeque, Elaine Fileman, Claudia Halsband, Rhys Goodhead, Julian Moger, and Tamara S. Galloway. 2013. Microplastic Ingestion by Zooplankton. *Environmental Science & Technology* 47: 6646–55.

Cullen, Jay T., and Maria T. Maldonado. 2013. Biogeochemistry of Cadmium and Its Release to the Environment. In *Cadmium: From Toxicity to Essentiality*. Edited by Astrid Sigel, Helmut Sigel and Roland K. O. Sigel. Dordrecht: Springer Netherlands, vol. 11, pp. 31–62.

Cunningham, Eoghan M., Sonja M. Ehlers, Jaimie T. A. Dick, Julia D. Sigwart, Katrin Linse, Jon J. Dick, and Konstadinos Kiriakoulakis. 2020. High Abundances of Microplastic Pollution in Deep-Sea Sediments: Evidence from Antarctica and the Southern Ocean. *Environmental Science & Technology* 54: 13661–71.

Dąbrowska, Jolanta, Marcin Sobota, Małgorzata Świąder, Paweł Borowski, Andrzej Moryl, Radosław Stodolak, Ewa Kucharczak, Zofia Zięba, and Jan K. Kazak. 2021. Marine Waste-Sources, Fate, Risks, Challenges and Research Needs. *International Journal of Environmental Research and Public Health* 18: 433. [CrossRef] [PubMed]

De Falco, Francesca, Maria Pia Gullo, Gennaro Gentile, Emilia Di Pace, Mariacristina Cocca, Laura Gelabert, and Marolda Brouta-Agnésa. 2018. Evaluation of Microplastic Release Caused by Textile Washing Processes of Synthetic Fabrics. *Environmental Pollution* 236: 916–25. [CrossRef] [PubMed]

de Lacerda, Luiz Drude, Felipe Goyanna, Moises F. Bezerra, and Aijanio G. B. Silva. 2017. Mercury Concentrations in Tuna (*Thunnus albacares* and *Thunnus obesus*) from the Brazilian Equatorial Atlantic Ocean. *Bulletin of Environmental Contamination and Toxicology* 98: 149–55. [CrossRef] de Lacerda, Luiz Drude, Rebecca Borges, and Alexander Cesar Ferreira. 2019. Neotropical mangroves: Conservation and sustainable use in a scenario of global climate change. *Aquatic Conservation: Marine and Freshwater Ecosystems* 29: 1347–64. [CrossRef]

de Lacerda, Luiz Drude, Rozane Valente Marins, and Francisco José da Silva Dias. 2020. Hg Inputs and Bioavailability to Global Climate Change in an Extreme Coastal Environment. *Frontiers in Earth Science* 8: 93. [CrossRef]

De Wit, Cynthia A., Rossana Bossi, Rune Dietz, Annekatrin Dreyer, Suzanne Faxneld, Svend Erik Garbus, Peter Hellström, Jan Koschorreck, Nina Lohmann, Anna Roos, and et al. 2020. Organohalogen compounds of emerging concern in Baltic Sea biota: Levels, biomagnification potential and comparisons with legacy contaminants. *Environment International* 144: 106037. [CrossRef] [PubMed]

De Witte, Bavo, Lisa Devriese, Karen Bekaert, Stefan Hoffman, Griet Vandermeersch, Kris Cooreman, and John Robbens. 2014. Quality Assessment of the Blue Mussel (*Mytilus edulis*): Comparison between Commercial and Wild Types. *Marine Pollution Bulletin* 85: 146–55. [CrossRef]

Desforges, Jean-Pierre W., Christian Sonne, Milton Levin, Ursula Siebert, Sylvain De Guise, and Rune Dietz. 2016. Immunotoxic effects of environmental pollutants in marine mammals. *Environmental International* 86: 126–39. [CrossRef] [PubMed]

Directive 2008/56. 2008. *Directive 2008/56 of the European Parliament and of the Council of 17 June 2008 establishing a framework for community action in the field of marine environmental policy (Marine Strategy Framework Directive).* Edited by European Union. Official Journal of the European Union L 164/19: Available online: https://eur-lex.europa.eu/legal-content/EN/TXT/PDF/?uri=CELEX:32008L0056&from=EN (accessed on 10 January 2021).

Directive 2013/39. 2013. *Directive 2013/39 of the European Parliament and of the Council of 12 August 2013, Amending Directives 2000/60/EC and 2008/105/EC as Regards Priority Substances in the Field of Water Policy.* Edited by European Union. Official Journal of the European Union L 226/1: Available online: https://eur-lex.europa.eu/LexUriServ/LexUriServ.do?uri=OJ:L:2013:226:0001:0017:EN:PDF (accessed on 10 January 2021).

Dixit, Ruchita, Wasiullah, Deepti Malaviya, Kuppusamy Pandiyan, Udai Singh, Asha Sahu, Renu Shukla, Bhanu P. Singh, Jai P. Rai, Pawan Kumar Sharma, and et al. 2015. Bioremediation of Heavy Metals from Soil and Aquatic Environment: An Overview of Principles and Criteria of Fundamental Processes. *Sustainability* 7: 2189–212. [CrossRef]

Eckert, Ester M., Andrea Di Cesare, Marie Therese Kettner, Maria Arias-Andres, Diego Fontaneto, Hans-Peter Grossart, and Gianluca Corno. 2018. Microplastics Increase Impact of Treated Wastewater on Freshwater Microbial Community. *Environmental Pollution* 234: 495–502. [CrossRef] [PubMed]

EFSA Scientific Committee. 2015. Statement on the Benefits of Fish/Seafood Consumption Compared to the Risks of Methylmercury in Fish/Seafood. *EFSA Journal* 13: 3982. [CrossRef]

Ehrlich, Paul R., David S. Dobkin, and Darryl Wheye. 1988. DDT and Birds. Available online: https://web.stanford.edu/group/stanfordbirds/text/essays/DDT_and_Birds.html (accessed on 19 January 2021).

Ekman, Elisabet, Gun Åkerman, Lennart Balk, and Leif Norrgren. 2004. Impact of PCB on Resistance to Flavobacterium Psychrophilum after Experimental Infection of Rainbow Trout Oncorhynchus Mykiss Eggs by Nanoinjection. *Diseases of Aquatic Organisms* 60: 31–39. [CrossRef] [PubMed]

El-Shahawi, Mohammad S. E.-S., Abdulaziz Hamza, Abdulaziz S. Bashammakh, and Wafaa T. Al-Saggaf. 2010. An Overview on the Accumulation, Distribution, Transformations, Toxicity and Analytical Methods for the Monitoring of Persistent Organic Pollutants. *Talanta* 80: 1587–97. [CrossRef] [PubMed]

Emmerton, Craig A., Jennifer A. Graydon, Jolie A. L. Gareis, Vincent L. St. Louis, Lance F. W. Lesack, Janelle K. A. Banack, Faye Hicks, and Jennifer Nafziger. 2013. Mercury Export to the Arctic Ocean from the Mackenzie River, Canada. *Environmental Science and Technology* 47: 7644–54. [CrossRef]

Eriksen, Markus, Laurent C. M. Lebreton, Henry S. Carson, Martin Thiel, Charles J. Moore, Jose C. Borerro, Francois Galgani, Peter G. Ryan, and Julia Reisser. 2014. Plastic pollution in the world's oceans: More than 5 trillion plastic pieces weighing over 250,000 tons afloat at sea. *PLoS ONE* 9: e111913. [CrossRef]

European Environment Agency. 2019. Trends in Emissions of Heavy Metals (EEA-33). Available online: https://www.eea.europa.eu/data-and-maps/daviz/emission-trends-of-heavy-metals-8#tab-chart_3.%C3%9F (accessed on 19 January 2021).

Falcó, Gemma, Juan M. Llobet, Ana Bocio, and José L. Domingo. 2006. Daily Intake of Arsenic, Cadmium, Mercury, and Lead by Consumption of Edible Marine Species. *Journal of Agricultural and Food Chemistry* 54: 6106–12. [CrossRef]

FDA. 2017. Mercury Levels in Commercial Fish and Shellfish (1990–2012). Available online: https://www.fda.gov/food/metals/mercury-levels-commercial-fish-and-shellfish-1990--2012 (accessed on 19 January 2021).

Ferriss, Bridget E., and Timothy E. Essington. 2011. Regional patterns in mercury and selenium concentrations of yellowfin tuna (*Thunnus albacares*) and bigeye tuna (*Thunnus obesus*) in the Pacific Ocean. *Canadian Journal of Fisheries and Aquatic Sciences* 68: 2046–56. [CrossRef]

Frias, João P. G. L., and Roisin Nash. 2019. Microplastics: Finding a consensus on the definition. *Marine Pollution Bulletin* 138: 145–47. [CrossRef] [PubMed]

Fry, D. Michael. 1995. Reproductive effects in birds exposed to pesticides and industrial chemicals. *Environmental Health Perspectives* 103 S7: 165–71.

Gago, Jorge, Olga Carretero, Ana V. Filgueiras, and Lucía Viñas. 2018. Synthetic Microfibers in the Marine Environment: A Review on Their Occurrence in Seawater and Sediments. *Marine Pollution Bulletin* 127: 365–76. [CrossRef] [PubMed]

Gall, Sarah C., and Richard C. Thompson. 2015. The impact of debris on marine life. *Marine Pollution Bulletin* 92: 170–79. [CrossRef]

Galligan, Thomas M., Brian C. Balmer, Lori H. Schwacke, Jennie L. Bolton, Brian M. Quigley, Patricia E. Rosel, Gina M. Ylitalo, and Ashley S.P. Boggs. 2019. Examining the relationships between blubber steroid hormones and persistent organic pollutants in common bottlenose dolphins. *Environmental Pollution* 249: 982–91. [CrossRef]

Garnaga, Galina, Eric Wyse, Sabine Azemard, Algirdas Stankevičius, and Stephen de Mora. 2006. Arsenic in Sediments from the Southeastern Baltic Sea. *Environmental Pollution* 144: 855–61. [CrossRef]

Gassel, Margy, and Chelsea M. Rochmann. 2019. The complex issue of chemicals and microplastic pollution: A case study in North Pacific lanternfish. *Environmental Pollution* 248: 1000–9. [CrossRef]

Gaw, Sally, Kevin V. Thomas, and Thomas H. Hutchinson. 2014. Sources, impacts and trends of pharmaceuticals in the marine and coastal environment. *Philosophical Transactions of the Royal Society* 369: 1656. [CrossRef] [PubMed]

GESAMP. 2016. Sources, Fate and Effects of Microplastics in the Marine Environment: A Global Assessment. In *Kershaw*. Edited by Peter J. Kershaw and C.M. (IMO/FAO/UNESCOIOC/UNIDO/WMO/IAEA/UN/UNEP/UNDP. Joint Group of Experts on The Scientific Aspects of Marine Environmental Protection) Rep. Stud. GESAMP. London: International Maritime Organization, No. 93. p. 220.

Gledhill, Martha, Aaron J. Beck, Beate Stamer, Christian Schlosser, and Eric P. Achterberg. 2019. Quantification of Munition Compounds in the Marine Environment by Solid Phase Extraction—Ultra High Performance Liquid Chromatography with Detection by Electrospray Ionisation—Mass Spectrometry. *Talanta* 200: 366–72. [CrossRef]

Gräslund, Sara, Katrin E. Holmström, and Ann Wahlström. 2003. A field survey of chemicals and biological products used in shrimp farming. *Marine Pollution Bulletin* 46: 81–90. [CrossRef]

Greiner, P., and I. Rönnefahrt. 2003. Management of environmental risks in the life cycle of pharmaceuticals. Paper presented at European Conference on Human and Veterinary Pharmaceuticals in the Environment (Envirapharma), Lyon, France, April 14–16.

GRID-Arendal. 2021. There Will Be More Plastic than Fish in the Oceans by 2050. Available online: http://marinelitter.no/myth3/ (accessed on 10 January 2021).

Grieb, Thomas M., Nicholas S. Fisher, Roxanne Karimi, and Leonard Levin. 2020. An assessment of temporal trends in mercury concentrations in fish. *Ecotoxicology* 29: 1739–49. [CrossRef]

Gust, Kurt A., Mitchell S. Wilbanks, Xin Guan, Mehdi Pirooznia, Tanwir Habib, Leslie Yoo, Henri Wintz, Chris D. Vulpe, and Edward J. Perkins. 2011. Investigations of Transcript Expression in Fathead Minnow (Pimephales Promelas) Brain Tissue Reveal Toxicological Impacts of RDX Exposure. *Aquatic Toxicology* 101: 135–45. [CrossRef] [PubMed]

Harrad, Stuart. 2010. Beyond the Stockholm Convention: An Introduction to Current Issues and Future Challenges in POPs Research. In *Persistent Organic Pollutants*. New York: Wiley, pp. 1–4.

Hartmann, Nanna B., Thorsten Hüffer, Richard C. Thompson, Martin Hassellöv, Anja Verschoor, Anders E. Daugaard, Sinja Rist, Therese Karlsson, Nicole Brennholt, Matthew Cole, and et al. 2019. Are We Speaking the Same Language? Recommendations for a Definition and Categorization Framework for Plastic Debris. *Environmental Science and Technology* 53: 1039–47. [CrossRef]

Helander, Björn, Mats Olsson, and Lars Reutergårdh. 1982. Residue levels of organochlorine and mercury compounds in unhatched eggs and the relationships to breeding success in white-tailed sea eagles *Haliaeetus albicilla* in Sweden. *Ecography* 5: 349–66. [CrossRef]

Helander, Björn, Anders Olsson, Anders Bignert, Lillemor Asplund, and Kerstin Litzén. 2002. The Role of DDE, PCB, Coplanar PCB and Eggshell Parameters for Reproduction in the White-tailed Sea Eagle (*Haliaeetus albicilla*) in Sweden. *Ambio* 31: 386–403. [CrossRef] [PubMed]

HELCOM. 1994. Report on Chemical Munitions Dumped in the Baltic Sea. Available online: https://helcom.fi/wp-content/uploads/2019/10/Report-on-chemical-munitions-dumped-in-the-Baltic-Sea.pdf (accessed on 19 January 2021).

HELCOM. 2010. Baltic Sea Environment Proceedings No. 120B. Hazardous Substances in the Baltic Sea. Available online: https://helcom.fi/media/publications/BSEP120B.pdf (accessed on 19 January 2021).

HELCOM. 2018. Metals (Lead, Cadmium and Mercury). Key Message. HELCOM Core Indicator Report. Available online: https://helcom.fi/media/core%20indicators/Metals-HELCOM-core-indicator-2018.pdf (accessed on 19 January 2021).

Hidalgo-Ruz, Valeria, Lars Gutow, Richard C. Thompson, and Martin Thiel. 2012. Microplastics in the Marine Environment: A Review of the Methods Used for Identification and Quantification. *Environmental Science & Technology* 46: 3060–75.

Hildebrandt, Lars, Marcus von der Au, Tristan Zimmermann, Anna Reese, Jannis Ludwig, and Daniel Pröfrock. 2020. A metrologically traceable protocol for the quantification of trace metals in different types of microplastic. *PLoS ONE* 15: e0236120. [CrossRef]

Hiltunen, Minna, Eeva-Riikka Vehniäinen, and Jussi V. K. Kukkonen. 2021. Interacting Effects of Simulated Eutrophication, Temperature Increase, and Microplastic Exposure on Daphnia. *Environmental Research* 192: 110304. [CrossRef] [PubMed]

Höher, Nicole, Raisa Turja, Matthias Brenner, Jenny Rattfelt Nyholm, Anders Östin, Per Leffler, Laura Butrimavičienė, Janina Baršienė, Mia Halme, Maaret Karjalainen, and et al. 2019. Toxic Effects of Chemical Warfare Agent Mixtures on the Mussel Mytilus Trossulus in the Baltic Sea: A Laboratory Exposure Study. *Marine Environmental Research* 145: 112–22. [CrossRef] [PubMed]

Hoydal, Katrin S., Bjarne Styrishave, Tomasz M. Ciesielski, Robert J. Letcher, Maria Dam, and Bjørn M. Jenssen. 2017. Steroid hormones and persistent organic pollutants in plasma from North-eastern Atlantic pilot whales. *Environmental Research* 195: 613–21. [CrossRef]

Huang, Wei, Biao Song, Jie Liang, Qiuya Niu, Guangming Zeng, Maocai Shen, Jiaqin Deng, Yuan Luo, Xiaofeng Wen, and Yafei Zhang. 2020. Microplastics and associated contaminants in the aquatic environment: A review on their ecotoxicological effects, trophic transfer, and potential impacts to human health. *Journal of Hazardous Materials* 405: 124187. [CrossRef] [PubMed]

Hulscher, Marlies E. J. L., Jos W. M. van der Meer, and Richard P. T. M. Grol. 2010. Antibiotic use: How to improve it? *International Journal of Medical Microbiology* 300: 351–56. [CrossRef]

Hung, Hayley, Athanasios A. Katsoyiannis, Eva Brorström-Lundén, Kristin Olafsdottir, Wenche Aas, Knut Breivik, Pernilla Bohlin-Nizzetto, Arni Sigurdsson, Hannele Hakola, Rossana Bossi, and et al. 2016. Temporal trends of Persistent Organic Pollutants (POPs) in arctic air: 20 years of monitoring under the Arctic Monitoring and Assessment Programme (AMAP). *Environmental Pollution* 217: 52–61. [CrossRef]

Hutz, Reinhold J., Michael J. Carvan III, Monika G. Baldridge, Lisa K. Conley, and Tisha King Heiden. 2006. Environmental Toxicants and Effects on Female Reproductive Function. *Trends in Reproductive Biology* 2: 1–11.

Imhof, Hannes K., Christian Laforsch, Alexandra C. Wiesheu, Johannes Schmid, Philipp M. Anger, Reinhard Niessner, and Natalia P. Ivleva. 2016. Pigments and Plastic in Limnetic Ecosystems: A Qualitative and Quantitative Study on Microparticles of Different Size Classes. *Water Research* 98: 64–74. [CrossRef] [PubMed]

Ivleva, Natalia P., Alexandra C. Wiesheu, and Reinhard Niessner. 2017. Microplastic in Aquatic Ecosystems. *Angewandte Chemie International Edition* 56: 1720–39. [CrossRef]

Jakimska, Anna, Piotr Konieczka, Krzysztof Skóra, and Jacek Namieśnik. 2011. Bioaccumulation of Metals in Tissues of Marine Animals, Part I: The Role and Impact of Heavy Metals on Organisms. *Polish Journal of Environmental Studies* 5: 1117–25.

Jambeck, Jenna R., Roland Geyer, Chris Wilcox, Theodore R. Siegler, Miriam Perryman, Anthony Andrady, Ramani Narayan, and Kara Lavender Law. 2015. Plastic waste inputs from land into the ocean. *Science* 347: 768–71. [CrossRef]

Jamieson, Alan J., Tamas Malkocs, Stuart B. Piertney, Toyonobu Fujii, and Zulin Zhang. 2017. Bioaccumulation of Persistent Organic Pollutants in the Deepest Ocean Fauna. *Nature Ecology & Evolution* 1: 0051.

Jones, Kevin C., and Pim de Voogt. 1999. Persistent Organic Pollutants (POPs): State of the Science. *Environmental Pollution* 100: 209–221. [CrossRef]

Juhasz, Albert, and Ravendra Naidu. 2007. Explosives: Fate, dynamics, and ecological impact in terrestrial and marine environments. *Reviews of Environmental Contamination and Toxicology* 191: 163–215. [PubMed]

Kelly, Barry C., Michael G. Ikonomou, Joel D. Blair, Anne E. Morin, and Frank A. P. C. Gobas. 2007. Food Web—Specific Biomagnification of Persistent Organic Pollutants. *Science* 317: 236. [CrossRef]

Khalid, Noreen, Muhammad Aqeel, Ali Noman, Mohamed Hashemd, Yasser S. Mostafa, Haifa Abdulaziz S. Alhaithloul, and Suliman M. Alghanem. 2021. Linking effects of microplastics to ecological impacts in marine environments. *Chemosphere* 44: 1901–8. [CrossRef]

Kinnberg, Karin, and Gunnar Toft. 2003. Effects of Estrogenic and Antiandrogenic Compounds on the Testis Structure of the Adult Guppy (*Poecilia reticulata*). *Ecotoxicology and Environmental Safety* 54: 16–24. [CrossRef]

Kirstein, Inga V., Sidika Kirmizi, Antje Wichels, Alexa Garin-Fernandez, Rene Erler, Martin Löder, and Gunnar Gerdts. 2016. Dangerous Hitchhikers? Evidence for Potentially *Pathogenic vibrio* spp. on Microplastic Particles. *Marine Environmental Research* 120: 1–8. [CrossRef]

Kirstein, Inga V., Fides Hensel, Alessio Gomiero, Lucian Iordachescu, Alvise Vianello, Hans B. Wittgren, and Jes Vollertsen. 2021. Drinking plastics?—Quantification and qualification of microplastics in drinking water distribution systems by µFTIR and Py-GCMS. *Water Research* 188: 116519. [CrossRef] [PubMed]

Kobusińska, Marta Ewelina, Krzysztof Konrad Lewandowski, Anna Panasiuk, Leszek Łęczyński, Magdalena Urbaniak, Tadeusz Ossowski, and Elżbieta Niemirycz. 2020. Precursors of Polychlorinated Dibenzo-p-Dioxins and Dibenzofurans in Arctic and Antarctic Marine Sediments: Environmental Concern in the Face of Climate Change. *Chemosphere* 260: 127605. [CrossRef] [PubMed]

Koelmans, Albert A., Adil Bakir, G. Allen Burton, and Colin R. Janssen. 2016. Microplastic as a vector for chemicals in the aquatic environment: Critical review and model-supported reinterpretation of empirical studies. *Environmental Science and Technology* 50: 3315–26. [CrossRef]

Kokalj, Anita J., Urban Kunej, and Tina Skalar. 2018. Screening study of four environmentally relevant microplastic pollutants: Uptake and effects on *Daphnia magna* and *Artemia franciscana*. *Chemosphere* 208: 522–29. [CrossRef]

Kümmerer, Klaus, ed. 2004. *Pharmaceuticals in the Environment*. Heidelberg: Springer.

Laist, David W. 1997. Impacts of Marine Debris: Entanglement of Marine Life in Marine Debris Including a Comprehensive List of Species with Entanglement and Ingestion Records. In *Marine Debris*. Edited by James M. Coe and Donald B. Rogers. New York: Springer New York, pp. 99–139.

Lammel, Gerhard, Andreas Heil, Irene Stemmler, Alice Dvorská, and Jana Klánová. 2013. On the Contribution of Biomass Burning to POPs (PAHs and PCDDs) in Air in Africa. *Environmental Science and Technology* 47: 11616–24. [CrossRef]

Landrigan, Philip J., John J. Stegeman, Lora E. Fleming, Denis Allemand, Donald M. Anderson, Lorraine C. Backer, Françoise Brucker-Davis, Nicolas Chevalier, Lilian Corra, Dorota Czerucka, and et al. 2020. Human Health and Ocean Pollution. *Annals of Global Health* 86: 1–64. [CrossRef]

Lebreton, Laurent, Boyan Slat, Francesco F. Ferrari, Bruno Sainte-Rose, Jack Aitken, Robert Marthouse, Sara Hajbane, Serena Cunsolo, Anna Schwarz, Aurore Levivier, and et al. 2018. Evidence That the Great Pacific Garbage Patch Is Rapidly Accumulating Plastic. *Scientific Reports* 8: 4666. [CrossRef] [PubMed]

Lerner, Darren T., Björn T. Björnsson, and Stephen D. McCormick. 2007. Effects of Aqueous Exposure to Polychlorinated Biphenyls (Aroclor 1254) on Physiology and Behavior of Smolt Development of Atlantic Salmon. *Aquatic Toxicology* 81: 329–36. [CrossRef] [PubMed]

Letcher, Robert J., Jan O. Bustnes, Rune Dietz, Bjørn M. Jenssen, Even H. Jørgensen, Christian Sonne, Jonathan Verreault, Mathilakath M. Vijayan, and Geir W. Gabrielsen. 2010. Exposure and effects assessment of persistent organohalogen contaminants in arctic wildlife and fish. *Science of The Total Environment* 408: 2995–3043. [CrossRef]

Li, W. C., H. F. Tse, and L. Fok. 2016. Plastic Waste in the Marine Environment: A Review of Sources, Occurrence and Effects. *Science of The Total Environment* 566–567: 333–49. [CrossRef]

Lischka, Alexandra, Thomas Lacoue-Labarthe, Hendrik Jan T. Hoving, Jamileh Javidpour, Jennifer L. Pannell, Véronique Merten, Carine Churlaud, and Paco Bustamante. 2018. High Cadmium and Mercury Concentrations in the Tissues of the Orange-Back Flying Squid, Sthenoteuthis Pteropus, from the Tropical Eastern Atlantic. *Ecotoxicology and Environmental Safety* 163: 323–30. [CrossRef]

Lithner, Delilah, Åke Larsson, and Göran Dave. 2011. Environmental and Health Hazard Ranking and Assessment of Plastic Polymers Based on Chemical Composition. *Science of The Total Environment* 409: 3309–24. [CrossRef]

Lohmann, Rainer. 2017. Microplastics Are Not Important for the Cycling and Bioaccumulation of Organic Pollutants in the Oceans-but Should Microplastics Be Considered POPs Themselves?: Should Microplastics Be Considered POPs. *Integrated Environmental Assessment and Management* 13: 460–65. [CrossRef] [PubMed]

López-Martínez, Sergio, Carmen Morales-Caselles, Julianna Kadar, and Marga L. Rivas. 2021. Overview of global status of plastic presence in marine vertebrates. *Global Change Biology* 27: 728–37. [CrossRef]

Lusher, Amy L., Matthew J. McHugh, and Richard C. Thompson. 2013. Occurrence of Microplastics in the Gastrointestinal Tract of Pelagic and Demersal Fish from the English Channel. *Marine Pollution Bulletin* 67: 94–99. [CrossRef]

Ma, Jianmin, Hayley Hung, Chongguo Tian, and Roland Kallenborn. 2011. Revolatilization of Persistent Organic Pollutants in the Arctic Induced by Climate Change. *Nature Climate Change* 1: 255–60. [CrossRef]

Ma, Jianmin, Hayley Hung, and Robie W. Macdonald. 2016. The Influence of Global Climate Change on the Environmental Fate of Persistent Organic Pollutants: A Review with Emphasis on the Northern Hemisphere and the Arctic as a Receptor. *Global and Planetary Change* 146: 89–108. [CrossRef]

Macdonald, Robie W., Tom J. Harner, and John C. Fyfe. 2005. Recent climate change in the Arctic and its impact on contaminant pathways and interpretation of temporal trend data. *Science of The Total Environment* 342: 5–86. [CrossRef] [PubMed]

Macgregor, Kenneth, Ian W. Oliver, Lynsay Harris, and Ian M. Ridgway. 2010. Persistent organic pollutants (PCB, DDT, HCH, HCB & BDE) in eels (*Anguilla anguilla*) in Scotland: Current levels and temporal trends. *Environmental Pollution* 158: 2402–11.

Magni, Stefano, François Gagné, Chantale André, Camilla Della Torre, Joëlle Auclair, Houda Hanana, Camilla Carla Parenti, Francesco Bonasoro, and Andrea Binelli. 2018. Evaluation of uptake and chronic toxicity of virgin polystyrene microbeads in freshwater zebra mussel *Dreissena polymorpha* (Mollusca: Bivalvia). *Science of The Total Environment* 631–632: 778–88. [CrossRef] [PubMed]

Mani, Thomas, Armin Hauk, Ulrich Walter, and Patricia Burkhardt-Holm. 2016. Microplastics Profile along the Rhine River. *Scientific Reports* 5: 17988. [CrossRef] [PubMed]

Mason, Ronald P., John R. Reinfelder, and François M. M. Morel. 1995. Bioaccumulation of Mercury and Methylmercury. *Water, Air, & Soil Pollution* 80: 915–21.

Mato, Yukie, Tomohiko Isobe, Hideshige Takada, Haruyuki Kanehiro, Chiyoko Ohtake, and Tsuguchika Kaminuma. 2001. Plastic Resin Pellets as a Transport Medium for Toxic Chemicals in the Marine Environment. *Environmental Science & Technology* 35: 318–24.

McGeer, James C., Kevin V. Brix, James M. Skeaff, David K. DeForest, Sarah I. Brigham, William J. Adams, and Andrew Green. 2003. Inverse Relationship between Bioconcentration Factor and Exposure Concentration for Metals: Implications for Hazard Assessment of Metals in the Aquatic Environment. *Environmental Toxicology and Chemistry* 22: 1017–37. [CrossRef]

Mendoza, Lorena M. Rios, Hrissi Karapanagioti, and Nancy Ramírez Álvarez. 2018. Micro(Nanoplastics) in the Marine Environment: *Current Knowledge and Gaps. Current Opinion in Environmental Science & Health* 1: 47–51.

Mintenig, Svenja M., Ivo Int-Veen, Martin G. J. Löder, Sebastian Primpke, and Gunnar Gerdts. 2017. Identification of Microplastic in Effluents of Waste Water Treatment Plants Using Focal Plane Array-Based Micro-Fourier-Transform Infrared Imaging. *Water Research* 108: 365–72. [CrossRef] [PubMed]

Missawi, Omayma, Noureddine Bousserrhine, Nesrine Zitouni, Maria Maisano, Iteb Boughattas, Giuseppe De Marco, Tiziana Cappello, Sabrina Belbekhouche, Mohamed Guerrouache, Vanessa Alphonse, and et al. 2021. Uptake, accumulation and associated cellular alterations of environmental samples of microplastics in the seaworm *Hediste diversicolor*. *Journal of Hazardous Materials* 406: 124287. [CrossRef]

Moos, Nadia von, Patricia Burkhardt-Holm, and Angela Köhler. 2012. Uptake and Effects of Microplastics on Cells and Tissue of the Blue Mussel *Mytilus edulis* L. after an Experimental Exposure. *Environmental Science and Technology* 46: 11327–35. [CrossRef]

Myers, Gary J., Philip W. Davidson, Gene E. Watson, Edwin van Wijngaarden, Sally W. Thurston, J. J. Strain, Conrad F. Shamlaye, and Pascal Bovet. 2015. Methylmercury Exposure and Developmental Neurotoxicity. *Bulletin of the World Health Organization* 93: 132A–2B. [CrossRef]

Nadal, Martí, Montse Marquès, Montse Mari, and José L. Domingo. 2015. Climate Change and Environmental Concentrations of POPs: A Review. *Environmental Research* 143: 177–85. [CrossRef]

Naik, Ravidas Krishna, Milind Mohan Naik, Priya Mallika D'Costa, and Fauzia Shaikh. 2019. Microplastics in ballast water as an emerging source and vector for harmful chemicals, antibiotics, metals, bacterial pathogens and HAB species: A potential risk to the marine environment and human health. *Marine Pollution Bulletin* 149: 110525. [CrossRef] [PubMed]

Neff, Jerry M. 1997. Ecotoxicology of Arsenic in the Marine Environment. *Environmental Toxicology and Chemistry* 16: 917–27. [CrossRef]

Neff, Jerry M. 2002. *Bioaccumulation in Marine Organisms: Effect of Contaminants from Oil Well Produced Water*. Amsterdam and London: Elsevier.

Niemikoski, Hanna, Martin Söderström, and Paula Vanninen. 2017. Detection of Chemical Warfare Agent-Related Phenylarsenic Compounds in Marine Biota Samples by LC-HESI/MS/MS. *Analytical Chemistry* 89: 11129–34. [CrossRef]

Nobre, Caio R., Beatriz B. Moreno, Aline V. Alves, Jonas de Lima Rosa, Heloisa da Rosa Franco, Denis Moledo de Souza Abessa, Luciane A. Maranho, Rodrigo B. Choueri, Paloma K. Gusso-Choueri, and Camilo D. S. Pereira. 2020. Effects of Microplastics Associated with Triclosan on the Oyster *Crassostrea brasiliana*: An Integrated Biomarker Approach. *Archives of Environmental Contamination and Toxicology* 79: 101–10. [CrossRef]

Nriagu, Jerome O., and Jozef M. Pacyna. 1988. Quantitative Assessment of Worldwide Contamination of Air, Water and Soils by Trace Metals. *Nature* 333: 134–39. [CrossRef] [PubMed]

Nyberg, Elisabeth, Suzanne Faxneld, Sara Danielsson, Ulla Eriksson, Aroha Miller, and Anders Bignert. 2015. Temporal and spatial trends of PCBs, DDTs, HCHs, and HCB in Swedish marine biota 1969–2012. *AMBIO* 44: 484–97. [CrossRef]

Obrist, Daniel, Yannick Agnan, Martin Jiskra, Christine L. Olson, Dominique P. Colegrove, Jacques Hueber, Christopher W. Moore, Jeroen E. Sonke, and Detlev Helmig. 2017. Tundra uptake of atmospheric elemental mercury drives Arctic mercury pollution. *Nature* 547: 201–4. [CrossRef] [PubMed]

Oliveira, Ana R., Andreia Sardinha-Silva, Paul L. R. Andrews, Dannielle Green, Gavan M. Cooke, Sarah Hall, Kirsty Blackburn, and António V. Sykes. 2020. Microplastics presence in cultured and wild-caught cuttlefish, *Sepia officinalis*. *Marine Pollution Bulletin* 160: 111553. [CrossRef] [PubMed]

Pandey, Deeptija, Ashwin Singh, Alagappan Ramanathan, and Manish Kumar. 2021. The combined exposure of microplastics and toxic contaminants in the floodplains of north India: A review. *Journal of Environmental Management* 279: 111557. [CrossRef]

Pavlaki, Maria D., Mário J. Araújo, Diogo N. Cardoso, Ana Rita R. Silva, Andreia Cruz, Sónia Mendo, Amadeu M. V. M. Soares, Ricardo Calado, and Susana Loureiro. 2016. Ecotoxicity and Genotoxicity of Cadmium in Different Marine Trophic Levels. *Environmental Pollution* 215: 203–12. [CrossRef] [PubMed]

Picó, Yolanda, and Damià Barceló. 2019. Analysis and Prevention of Microplastics Pollution in Water: Current Perspectives and Future Directions. *ACS Omega* 4: 6709–19. [CrossRef]

Plastics Europe. 2020. Plastics—The Facts 2020. Available online: https://www.plasticseurope. org/en/resources/publications/4312-plastics-facts-2020/ (accessed on 10 February 2021).

Plastics Europe. n.d. Annual Review 2017–2018. Available online: https://www.plasticseurope. org/en/ (accessed on 14 January 2021).

Porter, James W., James V. Barton, and Cecilia Torres. 2011. Ecological, Radiological, and Toxicological Effects of Naval Bombardment on the Coral Reefs of Isla de Vieques, Puerto Rico. In *Warfare Ecology*. Edited by Gary E. Machlis, Thor Hanson, Zdravko Špirić and Jean E. McKendry. Dordrecht: Springer, pp. 65–122.

Qiu, Xuchun, Souvannasing Saovany, Yuki Takai, Aimi Akasaka, Yoshiyuki Inoue, Naoaki Yakata, Yangqing Liu, Mami Waseda, Yohei Shimasaki, and Yuji Oshima. 2020. Quantifying the vector effects of polyethylene microplastics on the accumulation of anthracene to Japanese medaka (*Oryzias latipes*). *Aquatic Toxicology* 228: 105643. [CrossRef] [PubMed]

Rahman, M. Azizur, Hiroshi Hasegawa, and Richard Peter Lim. 2012. Bioaccumulation, Biotransformation and Trophic Transfer of Arsenic in the Aquatic Food Chain. *Environmental Research* 116: 118–35. [CrossRef]

Rheinheimer, Gerhard. 1998. Pollution in the Baltic Sea. *Naturwissenschaften* 85: 318–29. [CrossRef] [PubMed]

Rhodes, Christopher J. 2018. Plastic pollution and potential solutions. *Science Progress* 101: 207–60. [CrossRef]

Rigaud, Cyril, Catherine M. Couillard, Jocelyne Pellerin, Benoît Légaré, Patrice Gonzalez, and Peter V. Hodson. 2013. Relative Potency of PCB126 to TCDD for Sublethal Embryotoxicity in the Mummichog (*Fundulus heteroclitus*). *Aquatic Toxicology* 128–129: 203–14. [CrossRef]

Rigét, Frank, Katrin Vorkamp, Igor Eulaers, and Rune Dietz. 2020. Influence of Climate and Biological Variables on Temporal Trends of Persistent Organic Pollutants in Arctic Char and Ringed Seals from Greenland. *Environmental Science: Processes & Impacts* 22: 993–1005.

Rist, Sinja, Bethanie Carney Almroth, Nanna B. Hartmann, and Therese M. Karlsson. 2018. A critical perspective on early communications concerning human health aspects of microplastics. *Science of the Total Environment* 626: 720–26. [CrossRef]

Rochman, Chelsea M., Mark A. Browne, A. J. Underwood, Jan A. van Franeker, Richard C. Thompson, and Linda A Amaral-Zettler. 2016. The ecological impacts of marine debris: Unraveling the demonstrated evidence from what is perceived. *Ecology* 97: 302–12. [CrossRef]

Romeo, Teresa, Battaglia Pietro, Cristina Pedà, Pierpaolo Consoli, Franco Andaloro, and Maria Cristina Fossi. 2015. First evidence of presence of plastic debris in stomach of large pelagic fish in the Mediterranean Sea. *Marine Pollution Bulletin* 95: 358–61. [CrossRef] [PubMed]

Roos, Anna M., Britt-Marie V. M. Bäcklin, Björn O. Helander, Frank F. Rigét, and Ulla C. Eriksson. 2012. Improved reproductive success in otters (*Lutra lutra*), grey seals (*Halichoerus grypus*) and sea eagles (*Haliaeetus albicilla*) from Sweden in relation to concentrations of organochlorine contaminants. *Environmental Pollution* 170: 268–75. [CrossRef]

Rosen, Gunther, and Guilherme R. Lotufo. 2007. Toxicity of explosive compounds to the marine mussel, *Mytilus galloprovincialis*, in aqueous exposures. *Ecotoxicology and Environmental Safety* 68: 228–36. [CrossRef] [PubMed]

Rosen, Gunther, and Guilherme R. Lotufo. 2010. Fate and effects of Composition B in multispecies marine exposures. *Environmental Toxicology and Chemistry* 29: 1330–37. [CrossRef] [PubMed]

Ross, Peter S., Stephen Chastain, Ekaterina Vassilenko, Anahita Etemadifar, Sarah Zimmermann, Sarah-Ann Quesnel, and Jane Eert. 2021. Pervasive Distribution of Polyester Fibres in the Arctic Ocean Is Driven by Atlantic Inputs. *Nature Communications* 12: 106. [CrossRef]

Rusiecka, Dagmara, Martha Gledhill, Angela Milne, Eric P. Achterberg, Amber L. Annett, Sov Atkinson, Antony Birchill, Johannes Karstensen, Maeve Lohan, Clarisse Mariez, and et al. 2018. Anthropogenic Signatures of Lead in the Northeast Atlantic. *Geophysical Research Letters* 45: 2734–43. [CrossRef]

Schartup, Amina T., Prentiss H. Balcom, Anne L. Soerensen, Kathleen J. Gosnell, Ryan S. D. Calder, Robert P. Mason, and Elsie M. Sunderland. 2015. Freshwater discharges drive high levels of methylmercury in Arctic marine biota. *Proceedings of the National Academy of Sciences of the United States of America* 112: 11789–94. [CrossRef]

Schmidt, Britta, Christian Sonne, Dominik Nachtsheim, Peter Wohlsein, Sara Persson, Rune Dietz, and Ursula Siebert. 2020. Liver histopathology of Baltic grey seals (*Halichoerus grypus*) over three decades. *Environment International* 145: 106110. [CrossRef] [PubMed]

Scopetani, Costanza, Alessandra Cincinelli, Tania Martellini, Emilia Lombardini, Alice Ciofini, Alessia Fortunati, Vittorio Pasquali, Samuele Ciattinim, and Alberto Ugolini. 2018. Ingested microplastic as a two-way transporter for PBDEs in *Talitrus saltator*. *Environmental Research* 167: 410–17. [CrossRef] [PubMed]

Sharma, Madhu D., Anjana I. Elanjickal, Juili S. Mankar, and Reddithota J. Krupadam. 2020. Assessment of cancer risk of microplastics enriched with polycyclic aromatic hydrocarbons. *Journal of Hazardous Materials* 389: 122994. [CrossRef]

Shen, Maocai, Shujing Ye, Guangming Zeng, Yaxin Zhang, Lang Xing, Wangwang Tang, Xiaofeng Wen, and Shaoheng Liu. 2020. Can Microplastics Pose a Threat to Ocean Carbon Sequestration? *Marine Pollution Bulletin* 150: 110712. [CrossRef] [PubMed]

Sieber, Ramona, Delphine Kawecki, and Bernd Nowack. 2020. Dynamic probabilistic material flow analysis of rubber release from tires into the environment. *Environmental Pollution* 258: 113573. [CrossRef]

Sobek, Anna, Kristina L. Sundqvist, Anteneh T. Assefa, and Karin Wiberg. 2015. Baltic Sea sediment records: Unlikely near-future declines in PCBs and HCB. *Science of The Total Environment* 518–519: 8–15. [CrossRef]

Sonne, Christian, Ursula Siebert, Katharina Gonnsen, Jean-Pierre Desforges, Igor Eulaers, Sara Persson, Anna Roos, Britt-Marie Bäcklin, Kaarina Kauhala, Morten Tange Olsen, and et al. 2020. Health effects from contaminant exposure in Baltic Sea birds and marine mammals: A review. *Environment International* 139: 105725. [CrossRef]

Stellmann, Jeanne M., Steven D. Stellmann, Richard Christian, Tracy Weber, and Carrie Tomasallo. 2003. The extent and patterns of usage of Agent Orange and other herbicides in Vietnam. *Nature* 422: 681–87. [CrossRef]

Strehse, Jennifer S., Daniel Appel, Catharina Geist, Hans-Jörg Martin, and Edmund Maser. 2017. Biomonitoring of 2,4,6-Trinitrotoluene and Degradation Products in the Marine Environment with Transplanted Blue Mussels (*M. edulis*). *Toxicology* 390: 117–23. [CrossRef]

Sumpter, John P. 1995. Feminized Responses in Fish to Environmental Estrogens. *Toxicology Letters* 82–83: 737–42. [CrossRef]

Teng, Man, HaiJun Zhang, Qiang Fu, XianBo Lu, JiPing Chen, and FuSheng Wei. 2013. Irrigation-induced pollution of organochlorine pesticides and polychlorinated biphenyls in paddy field ecosystem of Liaohe River Plain, China. *Chinese Science Bulletin* 58: 1751–59. [CrossRef]

Teng, Jia, Jianmin Zhao, Xiaopeng Zhu, Encui Shan, Chen Zhang, Wenjing Zhang, and Qing Wang. 2021. Toxic effects of exposure to microplastics with environmentally relevant shapes and concentrations: Accumulation, energy metabolism and tissue damage in oyster Crassostrea gigas. *Environmental Pollution* 269: 116169. [CrossRef]

Uddin, Saif, Scott W. Fowler, and Montaha Behbehani. 2020. An assessment of microplastic inputs into the aquatic environment from wastewater streams. *Marine Pollution Bulletin* 160: 111538. [CrossRef] [PubMed]

Ullrich, Susanne M., Trevor W. Tanton, and Svetlana A. Abdrashitova. 2001. Mercury in the Aquatic Environment: A Review of Factors Affecting Methylation. *Critical Reviews in Environmental Science and Technology* 31: 241–93. [CrossRef]

United Nations Environment Programme. 2014. Plastic Waste Causes Financial Damage of US$13 Billion to Marine Ecosystems Each Year as Concern Grows over Microplastics. Available online: https://www.unenvironment.org/news-and-stories/press-release/plastic-waste-causes-financial-damage-us13-billion-marine-ecosystems (accessed on 14 January 2021).

United Nations Environment Programme. 2017. Clean Seas: More than 800 People Pledge to Stop Using Cosmetics Containing Microbeads. Available online: https://www.unenvironment.org/news-and-stories/story/cleanseas-more-800-people-pledge-stop-using-cosmetics-containing-microbeads (accessed on 14 January 2021).

United Nations Environment Programme. 2018a. Stockholm Convention on Persistent Organic Pollutants (POPS). Available online: http://chm.pops.int/Portals/0/download.aspx?d=UNEP-POPS-COP-CONVTEXT-2017.English.pdf (accessed on 19 January 2021).

United Nations Environment Programme. 2018b. Global Mercury Assessment 2018. Available online: https://www.unep.org/resources/publication/global-mercury-assessment-2018 (accessed on 19 January 2021).

United Nations Environment Programme. 2019. Minamata Convention on Mercury. Available online: http://www.mercuryconvention.org/Portals/11/documents/Booklets/COP3-version/Minamata-Convention-booklet-Sep2019-EN.pdf (accessed on 19 January 2021).

Vallius, Henry. 2014. Heavy Metal Concentrations in Sediment Cores from the Northern Baltic Sea: Declines over the Last Two Decades. *Marine Pollution Bulletin* 79: 359–64. [CrossRef]

Van Cauwenberghe, Lisbeth, Lisa Devriese, François Galgani, Johan Robbens, and Colin R. Janssen. 2015. Microplastics in Sediments: A Review of Techniques, Occurrence and Effects. *Particles in the Oceans: Implication for a Safe Marine Environment* 111: 5–17. [CrossRef] [PubMed]

Van Ginneken, Vincent, Arjan Palstra, Pim Leonards, Maaike Nieveen, Hans van den Berg, Gert Flik, Tom Spanings, Patrick Niemantsverdriet, Guido van den Thillart, and Albertinka Murk. 2009. PCBs and the Energy Cost of Migration in the European Eel (*Anguilla anguilla* L.). *Aquatic Toxicology* 92: 213–20. [CrossRef] [PubMed]

Van Sebille, Erik, Stefano Aliani, Kra L. Law, Nikolai Maximenko, José M. Alsina, Andrei Bagaev, Melanie Bergmann, Bertrand Chapron, Irina Chubarenko, Andrés Cózar, and et al. 2020. The physical oceanography of the transport of floating marine debris. *Environmental Research Letters* 15: 023003. [CrossRef]

Vasquez, Marlen I., A. Lambrianides, Manday Schneider, Klaus Kümmerer, and Despo Fatta-Kassinos. 2014. Environmental side effects of pharmaceutical cocktails: What we know and what we should know. *Journal of Hazardous Materials* 279: 169–89. [CrossRef]

Vassilopoulou, Loukia, Christos Psycharakis, Demetrios Petrakis, John Tsiaoussis, and Aristides M. Tsatsakis. 2017. Obesity, Persistent Organic Pollutants and Related Health Problems. *Advances in Experimental Medicine and Biology* 960: 81–110.

Vert, Michel, Yoshiharu Doi, Karl-Heinz Hellwich, Michael Hess, Philip Hodge, Przemyslaw Kubisa, Marguerite Rinaudo, and François Schué. 2012. Terminology for Biorelated Polymers and Applications (IUPAC Recommendations 2012). *Pure and Applied Chemistry* 84: 377–410. [CrossRef]

Vianello, Alvise, Alfredo Boldrin, Paolo Guerriero, Vanessa Moschino, Rocco Rella, Alberto Sturaro, and Luisa Da Ros. 2013. Microplastic Particles in Sediments of Lagoon of Venice, Italy: First Observations on Occurrence, Spatial Patterns and Identification. *Estuarine, Coastal and Shelf Science* 130: 54–61. [CrossRef]

Vitousek, Peter M., Harold A. Mooney, Jane Lubchenco, and Jerry M. Melillo. 1997. Human Domination of Earth's Ecosystems. *Science* 25: 494–99. [CrossRef]

Wang, Ying, Zheng Mao, Mingxing Zhang, Guanghui Ding, Jingxian Sun, Meijia Du, Quanbin Liu, Yi Cong, Fei Jin, Weiwei Zhang, and et al. 2019a. The uptake and elimination of polystyrene microplastics by the brine shrimp, Artemia parthenogenetica, and its impact on its feeding behavior and intestinal histology. *Chemosphere* 234: 123–31. [CrossRef] [PubMed]

Wang, Xiaoping, Chuanfei Wang, Tingting Zhu, Ping Gong, Jianjie Fu, and Zhiyuan Cong. 2019b. Persistent Organic Pollutants in the Polar Regions and the Tibetan Plateau: A Review of Current Knowledge and Future Prospects. *Environmental Pollution* 248: 191–208. [CrossRef] [PubMed]

Wang, Ting, Lin Wang, Qianqian Chen, Nicolas Kalogerakis, Rong Ji, and Yini Ma. 2020. Interactions between microplastics and organic pollutants: Effects on toxicity, bioaccumulation, degradation, and transport. *Science of The Total Environment* 748: 142427. [CrossRef]

Wani, Ab Latif, Anjum Ara, and Jawed Ahmad Usmani. 2015. Lead Toxicity: A Review. *Interdisciplinary Toxicology* 8: 55–64. [CrossRef]

Weber, Annkatrin, Nina Jeckel, and Martin Wagner. 2020. Combined Effects of Polystyrene Microplastics and Thermal Stress on the Freshwater Mussel Dreissena Polymorpha. *Science of The Total Environment* 718: 137253. [CrossRef] [PubMed]

WHO. 2003. *Elemental Mercury and Inorganic Mercury Compounds: Human Health Aspects. Concise International Chemical Assessment Document 50*. Geneva: WHO.

Woodall, Lucy C., Anna Sanchez-Vidal, Miquel Canals, Gordon L. J. Paterson, Rachel Coppock, Victoria Sleight, Antonio Calafat, Alex D. Rogers, Bhavani E. Narayanaswamy, and Richard C. Thompson. 2014. The Deep Sea Is a Major Sink for Microplastic Debris. *Royal Society Open Science* 1: 140317. [CrossRef]

World Economic Forum. 2016. *The New Plastics Economy: Rethinking the Future of Plastics*. Industry Agenda REF 080116. Cologny and Geneva: Industry Agenda.

Wright, Stephanie L., and Frank J. Kelly. 2017. Plastic and human health: A micro issue? *Environmental Science and Technology* 51: 6634–47. [CrossRef]

Wu, Wenzhong, Wen Li, Ying Xu, and Jianwei Wang. 2001. Long-Term Toxic Impact of 2,3,7,8-Tetrachlorodibenzo-p-Dioxin on the Reproduction, Sexual Differentiation, and Development of Different Life Stages of Gobiocypris Rarus and Daphnia Magna. *Ecotoxicology and Environmental Safety* 48: 293–300. [CrossRef]

Zantis, Laura J., Emma L. Carroll, Sarah E. Nelms, and Thijs Bosker. 2021. Marine mammals and microplastics: A systematic review and call for standardisation. *Environmental Pollution* 269: 116142. [CrossRef]

Zeilinger, Jana, Thomas Steger-Hartmann, Edmund Maser, Stephan Goller, Richardus Vonk, and Reinhard Länge. 2009. Effects of synthetic gestagens on fish reproduction. *Environmental Toxicology and Chemistry* 28: 2663. [CrossRef]

Zhang, Yanxu, Lyatt Jaeglé, LuAnne Thompson, and David G. Streets. 2014. Six Centuries of Changing Oceanic Mercury: Anthropogenic Mercury in Ocean. *Global Biogeochemical Cycles* 28: 1251–61. [CrossRef]

Zhang, Hanxiao, Shouliang Huo, Kevin M. Yeager, Chaocan Li, Beidou Xi, Jingtian Zhang, Zhuoshi He, and Chunzi Ma. 2019. Apparent Relationships between Anthropogenic Factors and Climate Change Indicators and POPs Deposition in a Lacustrine System. *Journal of Environmental Sciences* 83: 174–82. [CrossRef]

Zhang, Xiyang, Dingyu Luo, Ri-Qing Yu, Zhenhui Xie, Lei He, and Yuping Wu. 2021. Microplastics in the endangered Indo-Pacific humpback dolphins (*Sousa chinensis*) from the Pearl River Estuary, China. *Environmental Pollution* 270: 116057. [CrossRef] [PubMed]

Ziccardi, Linda M., Aaron Edgington, Karyn Hentz, Konrad J. Kulacki, and Susan K. Driscoll. 2016. Microplastics as vectors for bioaccumulation of hydrophobic organic chemicals in the marine environment: A state-of-the-science review. *Environmental Toxicology and Chemistry* 35: 1667–76. [CrossRef] [PubMed]

The Featuring of Small-Scale Fishers in SDG 14: Life below but Also above Water

Maarten Bavinck

1. Introduction

SDG 14, or 'life below water', addresses a rich variety of issues regarding the oceans. As the sustainable use of marine life depends very much on 'life above water'[1]—i.e., on human activities—its targets include people in various ways. Within its rich pallet of goals at the global level, target 14B stands out for its apparent pocket-size. It reads: "Provide access for small-scale artisanal fishers to marine resources and markets". From where did this goal and this category of people emerge? Moreover, where is target 14B leading us? These are the questions guiding this paper.

I first sketch the background leading to the inclusion of this goal in the SDGs. Attention then turns to small-scale fishers themselves and the sustainability challenges they face. Finally, I discuss the intentions with which this part of SDG 14 is currently being pursued.

2. Background

Marine fishers in the world are estimated to number thirty nine million and approximately 90% of them (or 35 million) pursue small-scale, or artisanal livelihoods (FAO 2020).[2] Actually, however, the numbers concerned are higher, as the small-scale sector also includes numerous processors, traders and other service providers, many of whom are women. Altogether, these professionals are sometimes grouped under the gender-neutral term 'fishworker'. Estimates of the total number of small-scale fishworkers in the world range from 107 million (Mills et al. 2011) to 200 million (Berkes et al. 2001). As we shall see below, most of them are based in Asia and Africa,

---

1 This is the title of Svein Jentoft's keynote address to the MARE People and the Sea Conference X on 25 June 2019 in Amsterdam.

2 The academic literature employs two terminologies for the same phenomenon: small-scale, or artisanal fishers. The former term suggests smallness of technology (boat, gear), capital and labor requirements and limitations in operational range. The latter term emphasizes the manual expertise involved and makes a contrast with industrialized fishing.

whereas in Europe, the Americas and Australia their numbers have been going down. Even in the latter regions, however, small-scale fishing continues, as various compendia demonstrate (Pascual-Fernández et al. 2020; Pinkerton and Davis 2015; Jentoft and Chuenpagdee 2015; Salas et al. 2019).

However, it is not so much the numbers of operators that has triggered the inclusion of Target 14B in the SDGs—far more important has been the adoption in 2014 by Food and Agriculture Organization (FAO) member countries of the Voluntary Guidelines for Securing Sustainable Small-Scale Fisheries (henceforth: the SSF Guidelines) and the political momentum that has come to surround the topic (Jentoft et al. 2017). The SSF Guidelines had a turbulent pre-history. Preparations can be said to have started with the FAO World Conference on Fisheries Management and Development (1984) in Rome,[3] which mobilized civil society and contributed to the rise of a global fisher movement, currently united under the flags of the World Fisher Forum (WFF) and the World Forum of Fisher Peoples (WFFP), as well as to a set of international NGOs, such as the International Collective in Support of Fishworkers (ICSF). Via parallel meetings, these civil society actors stood up to challenge the priority that governments were giving to the industrialization of fisheries and highlighted the problems that this was causing for the population of small-scale fishworkers that continued to inhabit the coastline. They argued that not only were industrialized fisheries provided unfair competition on inshore fishing grounds, these large-scale players were also causing substantial material damages to the small-scale fisheries sector. More fundamentally, these civil society actors noted that small-scale fisheries are not as old-fashioned as they are sometimes suggested to be. Instead, small-scale fisheries provide an alternative, and in many ways more appropriate model for realizing sustainability. After all, small- scale fisheries are frequently less fuel-intensive, more selective, less destructive of marine habitats, and socially and economically more inclusive. From this viewpoint, small-scale fishworkers, instead of being written off, require strong protection and sustenance. Technological subsidiarity is the direction to be pursued (Mathew 2005; Bavinck and Jentoft 2011).

To the rear of the dispute over appropriate modes of fishing lay another, more comprehensive transformation. By the 1970s, public opinion on capture fisheries and the future of the oceans had started to change. While the first three quarters of the twentieth century were dominated by the notion of an unlimited ocean and the

---

[3]   The organization of which followed from the United Nations Law of the Sea Convention that was concluded in 1982 (personal communication Sebastian Mathew).

creed of fisheries modernization, the planetary boundaries of the ocean were now becoming starkly apparent. 'Overfishing' had become the new concern, and policy efforts now came to focus on harmful fishing efforts. The FAO was called upon to develop guidelines for all capture fisheries.[4] The ultimate result of this effort was the Code of Conduct for Responsible Fisheries (1995).

The latter agreement, which is voluntary in nature, contains barely any reference to small-scale fishing. However, there, under Article 6 on general principles, is the following text:

> Recognizing the important contributions of artisanal and small-scale fisheries to employment, income and food security, States should appropriately protect the rights of fishers and fishworkers, particularly those engaged in subsistence, small-scale and artisanal fisheries, to a secure and just livelihood, as well as preferential access, where appropriate, to traditional fishing grounds and resources in the waters under their national jurisdiction (Article 6.18 of the Code of Conduct for Responsible Fisheries).

This article in the Code of Conduct underlines not only the contributions small-scale fishworkers are held to make but emphasizes their protection, suggesting that they even deserve preferential access to fishing grounds and resources.[5]

While proponents were pleased that the concerns of small-scale fishworkers were finally being recognized and included in international legislation, albeit of a 'soft' kind, a more specific, and elaborate instrument for the protection of the sector was felt to be essential. Efforts thus continued to realize a code specifically for the small-scale fisheries sector. This FAO-spearheaded endeavor involved many years of planning, extensive consultation with civil society actors, and intense negotiation among member states. One of the scholars partaking in this effort aptly summarizes the joy that accompanied the realization of the SSF Guidelines: "For the millions small-scale fishing people around the world, many of whom are poor and marginalized, this was no doubt an historic event and a potential turning point." (Jentoft 2014, pp. 1–2).

In line with other trends in international law, the SSF Guidelines follow a human rights approach. Article 3.1.1 in Part I reads:

---

[4]   The Code of Conduct for Responsible Fisheries (1995) originated in the Cancun Declaration (1992), which was adopted in response to the tuna-dolphin dispute between Mexico and the United States (personal communication Sebastian Mathew).

[5]   Needless to say, preferential access is more of a paper than a physical reality. While many governments pay obeisance to the notion of artisanal fishing zones, the implementation hereof is heavily flawed and deficient.

Recognizing the inherent dignity and the equal and inalienable human rights of all individuals, all parties should recognize, respect, promote and protect the human rights principles and their applicability to communities dependent on small-scale fisheries.

Its guiding principles thus emphasize the respect of cultures (Art. 3.1.2), non-discrimination (Art. 3.1.3), gender equality and equity (Art. 3.1.4), as couched in justice and fair treatment of all people and peoples (Art. 3.1.5), "ensuring active, free, effective, meaningful and informed participation" for small-scale fishers in relevant decision-making processes (Art. 3.1.6). The latter section of the SSF Guidelines identifies key topics of relevance to small-scale fishers: the securement of tenure rights to resources (Art. 5), social development, employment and decent work (Art. 6), value chains, post-harvest and trade (Art. 7), gender equality (Art. 8), and disaster risks and climate change (Art. 9).

While the adoption of the SSF Guidelines was widely celebrated, their proponents recognized, however, that, as in the adage, 'the proof of the pudding is in the eating'. As Jentoft puts it: "the challenge now is to make sure that [the SSF Guidelines] will be implemented" (2014, p. 1). A broad coalition of international actors—including intergovernmental organizations (FAO), international NGOs (e.g. the International Collective for the Support of Fishworkers—ICSF), research organizations (e.g., WorldFish/The Transnational Institute—TNI) and networks (e.g., Too Big To Ignore—TBTI), and member organizations (e.g., WFF and WFFP)—are engaged in precisely this effort, and it is their perseverance that has probably triggered the inclusion of the small-scale fishworker issue in SDG 14.[6]

3. The Condition of Small-Scale Fisheries

I noted above the fact that—according to criteria of technology, as well as labor and capital extensiveness—the large majority of the world's fisher peoples are small in scale. Still, their variety is impressive. As Jentoft and Chuenpagdee (2015) point out: "Globally small-scale fisheries display enormous diversity as a result of differences in natural, social, cultural and political factors" (p. 37). This diversity makes what is considered small-scale in some contexts (North America, Europe)

---

[6] Thus, Sebastian Mathew points out that ICSF worked with the Brazilian and the Indian delegations to retain Article 24(2)(b) of the United Nations Fish Stocks Agreement (1995) in the review conference outcome document. ICSF also worked with the Brazilian and EU delegations to retain this reference in the outcome document of the Rio+20 Ocean Conference in 2012. From this document, it moved into the SDG 14 (personal communication).

large-scale in others (Asia/Africa/Latin America). As a consequence, there exists no single definition of what constitutes small-scale fishing. Some governing actors, like the European Commission, have formulated a minimalist description, limiting the field of small-scale fisheries to those making use of vessels less than 12 m in length. However, as Pascual-Fernández et al. (2020) in their chapter on the small-scale fisheries of Europe point out, even this definition is problematic, excluding some fishers that properly belong, and including some that should not be included in the category.

Moreover, small-scale fishing is never fixed but highly dynamic: target species, fishing grounds, fishing practice, and market changes from one moment to the next. What is common practice today may change tomorrow. As a consequence of these dynamics, Johnson (2006) suggests that fishers should globally be arranged on a fluid scale ranging from subsistence to industrial fisheries. Jentoft and Chuenpagdee (2015) take a different approach, however, arguing that small-scale fisheries "must always be considered in their particular context" (p. 37).

Table 1 provides a summary overview of fisher numbers by continent. Although these figures—which relate only to those actually engaged in the catching of fish—do not distinguish small-scale from industrial-scale fishers, we can safely assume that, in all cases, the large majority of fishers belong to the ranks of small-scale operators.

Table 1. Fisher numbers by geographical region in thousands (2018).

Region	Numbers	Change Since 1995 (%)
Africa	5021	+183%
Asia	30,768	+127%
Europe	272	−39%
The Americas	2455	+137%
Oceania	460	+0%
Total	38,976	+132%

Source: FAO (2020), Table 12.

Table 1 points out that in some regions of the world, the fisher population has increased substantially, whereas in others, their numbers have declined. Such variations relate to differing conditions for social mobility. Whereas in the Global South, macro-economic and social conditions tend to keep people locked in fisheries, a very different set of conditions in the Global North has encouraged them to seek other forms of employment. It is not unreasonable to assume that in some situations people are in fishing largely 'because they are poor' (Béné 2003) and do not have other livelihood opportunities. In an earlier publication related to South India, I have noted that a substantial immigration of poor agriculturalists into fishing had probably occurred (Bavinck 2011). Such a process appears to have taken place in other parts of the world too.

4. Looking Ahead

I have argued that international attention for small-scale fishing has increased dramatically in recent years, with some arguing that it is now just 'too big to ignore'.[7] At the same time, one cannot ignore the dark clouds that gather on the horizon. While fishing is one of humankind's oldest maritime occupations, and the sector still provides large numbers of people with gainful employment, the range of competing endeavors, gathered under the denominators of 'blue economy' or 'blue growth', is increasing fast. Aquaculture, coastal tourism, deep sea mining, oil and gas exploitation and wind parks are making their mark on oceanic and coastal space, resulting in complaints of 'ocean grab' (Bennett et al. 2015) and 'coastal grab' (Bavinck et al. 2017). With governments anxious to capitalize on new investment opportunities and opting to see small-scale fisheries as an obsolete enterprise, small-scale fishworkers suffer stiff competition for shoreline and ocean space. Some knowledgeable authors (Percy and O'Riordan 2020) therefore express pessimism toward the future.

The international conservation movement has been making demands of small-scale fisheries, pointing out that the erosion of biodiversity and the challenges of climate change require urgent addressal (Garcia et al. 2014). Although small-scale fisheries arguably have a better environmental record than large-scale fisheries (Kolding et al. 2014), small-scale fishworkers also have a contribution to make toward achieving resilience. The objective, then, as argued by Charles et al. (2014), is to achieve win-win-win solutions across the three dimensions of sustainable development, while simultaneously taking into account the specific vulnerabilities of small-scale fisheries.

The fact that small-scale fisheries are included in SDG 14 is a sign that resilience is still deficient. Target 14B emphasizes provision of access to marine resources as well as to markets. The former can be achieved in various ways: by delimiting

---

[7] Too Big to Ignore (TBTI) is the title of a recent project funded by the Canadian Social Science Research Council. This project has contributed to a flood of new academic interest in the topic.

specific small-scale fishing zones, providing quota, or, to the contrary, by limiting the prerogatives of other actors, such as industrial fishers. Access to markets is a parallel concern, meant to make small-scale fisheries economically sustainable, meanwhile contributing to the food security of rural and urban populations. Both forms of access are crucial to the future of small-scale fisheries. In turn, small-scale fishworkers have responsibilities toward other targets of SDG 14, such as the sustainable protection and management of marine and coastal ecosystems (Article 14.2) and the effective regulation of harvesting (Article 14.4).

In past decades, small-scale fishworkers have garnered a wealth of influential support, and their cause is being fought in international, national and local arenas. The alliances they have forged with other rural movements, such as the Via Campesina, have lent force to their actions. The oncoming International Year of Artisanal Fisheries and Aquaculture (2022) will provide a useful rallying point for continued engagement, with, as a provisional objective, the realization of Target 14B in 2030. Time will tell of its accomplishment.

5. Conclusions

This chapter has attempted to explain why SDG 14, which largely deals with 'life below water', includes a direct reference to the human dimension. Article 14B expresses support of the livelihoods of small-scale fishers with regard to access to resources and markets. I have argued that the inclusion of this anomaly is rooted in the international history of fisheries regulation, and the growing attention that social movements have drawn to the marginalization of small-scale fishworkers. The last section looked forward to the future of this sub-sector, which was sketched as highly uncertain. However, the fact that small-scale fishers have organized themselves provides some confidence that Article 14B may actually have teeth.

Funding: This research received no external funding.

Acknowledgments: I gratefully acknowledge the comments made by Svein Jentoft and Sebastian Mathew on an earlier version of this paper, as well as the reviews provided by two anonymous reviewers.

Conflicts of Interest: The author is a maritime geographer with sympathies for the small-scale fisher movement, as expressed by his current chairpersonship of the International Collective for the Support of Small-Scale Fishworkers.

References

Bavinck, Maarten. 2011. Wealth, poverty, and immigration—The role of institutions in the fisheries of Tamil Nadu, India. In *Poverty Mosaics: Realities and Prospects in Small-Scale Fisheries*. Edited by Svein Jentoft and Arne Eide. Dordrecht: Springer, pp. 173–91.

Bavinck, Maarten, and Svein Jentoft. 2011. Subsidiarity as a guiding principle for small-scale fishing. In *World Small-Scale Fisheries: Contemporary Visions*. Edited by Ratana Chuenpagdee. Delft: Eburon Academic Publishers, pp. 311–21.

Bavinck, Maarten, Fikret Berkes, Anthony Charles, Ana Carolina Esteves Dias, Nancy Doubleday, Prateep Nayak, and Merle Sowman. 2017. The impact of coastal grabbing on community conservation—A global reconnaissance. *Maritime Studies (MAST)* 16: 8. [CrossRef]

Béné, Christophe. 2003. When fisheries rhymes with poverty: A first step beyond the old paradigm on poverty in small-scale fisheries. *World Development* 31: 949–75. [CrossRef]

Bennett, Nathan James, Hugh Govan, and Terre Satterfield. 2015. Ocean-grabbing. *Marine Policy* 57: 61–68. [CrossRef]

Berkes, Fikret, Robin Mahon, Patrick McConney, Richard Pollnac, and Robert Pomeroy, eds. 2001. *Managing Small-Scale Fisheries: Alternative Directions and Methods*. Ottawa: International Development Research Centre.

Charles, Anthony, Serge M. Garcia, and Jake Rice, eds. 2014. A tale of two streams: Synthesizing governance of marine fisheries and biodiversity conservation. In *Governance for Fisheries and Marine Conservation: Interactions and Co-Evolution*. Hoboken: Wiley-Blackwell.

Food and Agriculture Organization (FAO). 2020. *State of World Fisheries and Aquaculture 2020*. Rome: FAO.

Garcia, Serge M., Jake Rice, and Anthony Charles, eds. 2014. Governance of marine fisheries and biodiversity conservation: convergence or coevolution. In *Governance for Fisheries and Marine Conservation: Interactions and Co-Evolution*. Hoboken: Wiley-Blackwell.

Jentoft, Svein. 2014. Walking the talk: Implementing the international voluntary guidelines for securing sustainable small-scale fisheries. *Maritime Studies* 13: 16. [CrossRef]

Jentoft, Svein, and Ratana Chuenpagdee, eds. 2015. Assessing governability of small-scale fisheries. In *Interactive Governance for Small-Scale Fisheries*. MARE Publication Series; Dordrecht: Springer, vol. 13.

Jentoft, Svein, Ratana Chuenpagdee, María José Barragán-Paladines, and Nicole Franz, eds. 2017. *The Small-Scale Fisheries Guidelines*. MARE Publication Series; Dordrecht: Springer, vol. 13.

Johnson, Derek Stephen. 2006. Category, narrative and value in the governance of small-scale fisheries. *Marine Policy* 30: 747–56. [CrossRef]

Kolding, Jeppe, Chris Bené, and Maarten Bavinck. 2014. Small-scale fisheries—Importance, vulnerability, and deficient knowledge. In *Governance for Fisheries and Marine Conservation: Interactions and Co-Evolution*. Edited by Serge M. Garcia, Jake Rice and Anthony Charles. Hoboken: Wiley-Blackwell.

Mathew, Sebastian. 2005. Fisheries and Their Contribution to Sustainable Development. Paper presented at the 6th Meeting of the United Nations Open-Ended Informal Consultative Process on Oceans and the Law of the Sea, New York, NY, USA, June 6–10.

Mills, David J., Lena Westlund, Gertjan de Graaf, Rolf Willmann, Yumiko Kura, and Kevin Kelleher. 2011. Underreported and undervalued: Small-scale fisheries in the developing world. In *Small-Scale Fisheries Management: Frameworks and Approaches for the Developing World*. Edited by Neil L. Andrew and Robert Pomeroy. Wallingford: CABI, pp. 1–15.

Pascual-Fernández, José J., Cristina Pita, and Maarten Bavinck, eds. 2020. Small-Scale Fisheries Take Centre-Stage in Europe (Once Again). In *Small-Scale Fisheries in Europe: Status, Resilience and Governance*. MARE Publication Series; Dordrecht: Springer, vol. 23.

Percy, Jeremy, and Brian O'Riordan. 2020. The EU Common Fisheries Policy and Small-Scale Fisheries: A Forgotten Fleet Fighting for Recognition. In *Small-Scale Fisheries in Europe: Status, Resilience and Governance*. Edited by José J. Pascual-Fernández, Cristina Pita and Maarten Bavinck. Cham: Springer.

Pinkerton, Evelyn, and Reade Davis. 2015. Neoliberalism and the politics of enclosure in North American small-scale fisheries. *Marine Policy* 61: 303–12. [CrossRef]

Salas, Silvia, Maria José Barragan-Paladines, and Ratana Chuenpagdee, eds. 2019. *Viability and Sustainability of Small-Scale Fisheries in Latin America and The Caribbean*. Dordrecht: Springer.

Global Processes in Ocean Policy:
An Opportunity to Create Coherence in
Governance Frameworks and Support the
Achievement of Conservation Goals

Ben Boteler, Carole Durussel, Sebastian Unger and Torsten Thiele

1. Introduction: Marine Biodiversity, Ecological Connectivity, and Global Processes for Conservation

The ocean is essential to all life on the planet. It covers more than 70% of the earth's surface and regulates the climate, provides essential resources and ecosystem services, hosts immense biodiversity and underpins human activities, such as fisheries, offshore oil and gas, and international trade, as well as recreational, educational and cultural activities (Wright et al. 2017). Pressure on marine biodiversity is largely caused by increasing human activities such as fishing and shipping, but also coastal and land-based activities such as oil and gas extraction, port development, agriculture, industry, urban expansion and tourism (Wright et al. 2017). Emerging activities such as deep seabed mining have the potential to cause further impacts on the marine environment in the future (Boteler et al. 2019a). Pressures from human activities include, amongst others, extraction of living species, physical disturbance to and destruction of the seabed, pollution from land and sea, and underwater noise and light (Boteler et al. 2019a). Compounding effects due to increases in anthropogenic CO_2 emissions have resulted in rising ocean acidity, declining oxygen levels, warming waters and shifting ocean currents (Boteler et al. 2019a). The recent reports from the Intergovernmental Science-Policy Platform on Biodiversity and Ecosystem Services (IPBES 2019) and the Intergovernmental Panel on Climate Change (IPCC 2019) confirm that ocean health continues to degrade, including from climate change, and necessitates increased efforts from states to protect and sustainably manage marine ecosystems (Boteler et al. 2019b).

The ocean is legally divided into different zones. States can declare marine areas of up to 200 nautical miles (based on determining a national baseline) as national jurisdiction comprising the territorial sea, contiguous zone or the exclusive economic zone (EEZ). Not all states exercise this right. Marine areas beyond 200 nautical miles are known as areas beyond national jurisdiction (ABNJ) and include both the water column and the seabed beyond national jurisdiction, known as the high seas and the area, respectively. They are legally distinct and states can claim extended entitlements over the continental shelf, meaning that there is less of the area than the high seas in ABNJ. While ABNJ and waters under national jurisdiction are legally established as distinct entities, they are highly connected ecologically. The same also applies to the high seas and the area. Hence, pollution, overfishing, mining or geoengineering experiments in the high seas and/or the area can result in ecological and socioeconomic impacts in coastal waters or the water column—and vice versa.

Ecological connectivity, both vertically within the water column and horizontally across ocean basins, is due to two factors: First, small marine organisms such as plankton and larvae, which cannot actively swim in the water column, and pollution, such as plastics, ghost fishing gears or oil, are transported through passive connectivity within the water column by ocean currents (Dunn et al. 2019; Popova et al. 2019). The strength and direction of ocean currents influence the temporal scale by which impacts from human activities may be identified or realized, ranging from within a few weeks to months, or even years, depending on the location of the impact (Boteler et al. 2019a). As ocean circulation shifts due to changes in seasonal, inter-annual and multi-decadal climate patterns, this can in turn affect e.g., the distribution of plankton or the location of upwelling and downwelling areas. In severe cases, this can result in a shift in species range and ultimately can affect marine ecosystems (Boteler et al. 2019a). Second, the active movement of marine species within the water column and across ocean basins, such as between feeding and breeding grounds, is recognized as active or migratory connectivity (Dunn et al. 2019; Popova et al. 2019). Many migratory species cross vast distances and straddle the boundaries between ABNJ and national waters in their life cycle, thereby connecting distant ecosystems (Dunn et al. 2019; Popova et al. 2019). Many of these species will also spend different stages of their lives (e.g., larval and adult) within different areas, with timescales ranging from a few hours to days or months (Di Franco et al. 2012; Dunn et al. 2019; Popova et al. 2019; Rogers et al. 2019). To be effective, ocean conservation efforts must consider both passive and active ecological connectivity, as well as between ocean basins (Dunn et al. 2019).

The existing ocean governance structure to sustainably manage human uses on and in the ocean and ensure conservation of marine species and ecosystems is fragmented, has legal and institutional gaps, and lacks full implementation and enforcement (Durussel et al. 2018; Gjerde et al. 2018). There is currently no comprehensive approach or coherent structure to bring together the legal, institutional or policy framework established for ocean conservation. The 1982 United Nations

Convention on the Law of the Sea (UNCLOS) provides for rules governing uses of the ocean and its resources, including ABNJ, and is considered the umbrella convention for the protection of the marine environment and sustainable use of ocean resources (UNGA 1992). However, these rules are limited and do not specify how states should conserve and sustainably use biodiversity in ABNJ. An uneven governance framework was created through the numerous regional and sectoral agreements, covering sectors such as fisheries, shipping and others adopted independently, both before and after UNCLOS came into force in 1994 (Durussel et al. 2018; Gjerde et al. 2018). For these reasons, the current ocean governance framework does not address the cumulative impacts placed on the marine environment due to human activities. Compounding this, numerous practical challenges also exist. For example, it is inherently difficult to convince institutions to cooperate on shared challenges or goals, and there is a general reluctance from states to commit funds on a sustained and sustainable basis to promote ocean governance as a priority within and across institutions. At the same time, not all institutions or actors across the ocean governance framework may be prepared to address or even be aware of global conservation goals, or coordinate to actively achieve and co-implement management measures (e.g., through data and knowledge exchange), or implement common sustainability principles, such as the precautionary principle, ecosystem approach, or participatory decision-making processes (Boteler et al. 2019b). Such lack of coordination also exists between the various national government agencies, further exacerbating challenges of ocean management and conservation. Hence, strengthening ocean governance at all levels and across all actors will be necessary to achieve global conservation goals.

Building on political momentum, three major global processes are currently underway in regard to global ocean governance under the umbrella of the United Nations. First, the development of an international legally binding agreement under UNCLOS for the conservation and sustainable use of marine biodiversity in areas beyond national jurisdiction (BBNJ) is being negotiated (from here on referred to as BBNJ process) (UNGA/RES/69/292 2015). Second, under the Convention on Biological Diversity (CBD), 20 Aichi Biodiversity Targets were adopted in 2010 as part of the CBD Strategic Plan for Biodiversity 2011–2020, in an effort to reduce pressures on biodiversity, promote its sustainable use and safeguard ecosystem functions (from here on referred to as CBD process). These policy targets are currently being discussed under the umbrella of the CBD and will lead to the development of updated and new biodiversity targets that will be adopted at the 15th Conference of the Parties (COP) to the CBD in October 2020 as part of the Post-2020 Global Biodiversity Framework (CBD/COP/DEC/14/34 2018). Third, the United Nations 2030 Agenda for Sustainable

Development, which focuses on 17 Sustainable Development Goals (SDG), including SDG 1 on the ocean, coasts and marine resources, aims to holistically address current global challenges to sustainability, including those specifically negatively affecting the oceans and their ecosystems (from here on referred to as SDG process) (UNGA 2015). Momentum for SDG 14 implementation has been triggered in particular through the 2017 UN Ocean Conference and voluntary commitments for ocean action by states and other actors (Neumann and Unger 2019). Despite these efforts and progress towards global conservation goals, marine biodiversity and ocean health continue to decline (IPBES 2019). Taking into account the BBNJ, SDG and CBD processes, this chapter highlights the need to ensure coherence across these global processes for marine conservation, and provides ways in which ocean governance can be strengthened to support global processes and marine conservation goals.

2. Understanding Global Processes for Marine Conservation

While specifics of the BBNJ, CBD and SDG processes differ, ultimately they have common overarching objectives to reduce the negative impacts from human activities on the marine environment and to ensure the conservation of marine ecosystems and sustainable use of marine resources. The most important difference is that the BBNJ process will create a legally binding global instrument which will establish rules and guidelines for how humans interact with ABNJ, while the post-2020 biodiversity framework and SDG processes set out policy targets and goals to guide state and societal actions. Viewed together, the BBNJ process, the Aichi Targets and the SDGs present an important opportunity for states to strengthen the overall ocean governance framework both globally and at the regional level, and thereby contribute to sustainable development and economic growth. Considering ecological connectivity, it is essential to consider conservation efforts and sustainable management of human activities, both within and beyond national jurisdiction. Particularly, strengthened collaboration and cooperation between global, regional (i.e., marine regions) and sectoral organizations will be needed to boost ocean governance efforts and will be an important step towards underpinning actions for the conservation and sustainable use of BBNJ, the targets of the Post-2020 Global Biodiversity Framework, and the SDGs.

2.1. Understanding the Scope and Nature of the BBNJ Negotiations, the Targets of the Post-2020 Global Biodiversity Framework, and the SDGs

The BBNJ process is a political process currently underway to negotiate an international legally binding agreement under UNCLOS on the conservation and sustainable use of BBNJ. After a decade of discussions in a working group, the United Nations General Assembly (UNGA) decided in 2015 to begin negotiating a BBNJ Agreement. A Preparatory Committee was established to make recommendations to the UNGA on the elements of a draft text and, since 2018 through the Intergovernmental Conference (IGC), to elaborate the text of the agreement (UNGA/RES/72/249 2018). Four elements provide the structure for negotiations (UNGA/RES/69/292 2015) and are:

- marine genetic resources (MGRs), including questions of their access and sharing of their benefits;
- area-based management tools (ABMTs), including marine protected areas (MPAs);
- environmental impact assessments (EIAs);
- and capacity building and the transfer of marine technology.

The effective implementation of the BBNJ Agreement will offer an opportunity to improve coordination between and among existing global and regional institutions. However, to do this will require a clear and coherent legal and institutional framework, both within marine regions (i.e., multiple states with a common interest in a specific marine ecosystem) and at the global level with regard to managing sectoral activities (Gjerde et al. 2018; Gjerde and Wright 2019). Although the BBNJ negotiations are markedly narrower in scope than the CBD and SDG processes, the legally binding nature of the future BBNJ Agreement makes this process much more politically sensitive than political declarations achieved under the CBD and SDG processes. Indeed, whereas the BBNJ process is still in the negotiation phase, it is expected to have some mechanism by which to enforce the agreed upon obligations, ensuring that states, or activities under state flags, adhere to the agreement.

Under the CBD, 20 Aichi Targets were adopted in 2010 by which states commit themselves to take action towards reaching specific biodiversity related objectives (UNEP/CBD/COP/DEC/X/2 2010). However, unlike the BBNJ Agreement, these are policy targets that create no legal obligations. Parties to the CBD submit National Biodiversity Strategies and Action Plans (NBSAPs) for the conservation and sustainable use of biodiversity or adapt existing national strategies or plans to reflect the objectives of the CBD. They also have to integrate, as far as possible and as appropriate, the conservation and sustainable use of biodiversity into relevant sectoral or cross-sectoral plans, programs and policies (UNGA 2015). The Aichi Targets, many of which are relevant to marine and coastal biodiversity, are reflected in the SDGs, and many of them are set for 2020. Of particular note, CBD Aichi Target 11 establishes that 10% of coastal and marine areas are conserved through ecologically representative and well-connected systems of protected areas and other effective area-based conservation measures by 2020. Compatible goals in regional strategies and policy objectives reflect this global goal. For example, this can be seen in the MPA network designated in the Baltic Sea region (HELCOM 2016). Indeed, efforts both in EEZs and ABNJ have been made to establish marine protected areas. These Aichi targets are currently being reviewed and it is expected that updated and possibly more ambitious, as well as new biodiversity targets, will be adopted at the upcoming CBD COP in 2020, as part of the Post-2020 Global Biodiversity Framework (CBD/COP/DEC/14/34 2018).

The UNGA adopted, in 2015, Resolution 70/01 on the 2030 Agenda for Sustainable Development, which sets out a global 'plan of action for people, planet and prosperity'. The 2030 Agenda puts forward a set of 17 globally applicable SDGs with 169 underlying targets (UNGA 2015). These goals take into consideration the need for economic, social, and environmental sustainability, and thus include a wide range of aspirations, from conservation and protection, to sustainable modes of production and consumption to peaceful and inclusive societies (UNGA 2015). SDG 14 is explicitly dedicated to the conservation and sustainable use of the oceans, seas and marine resources for sustainable development. The 10 targets set in SDG 14 mostly reflect existing policy agreements, such as the 2002 World Summit on Sustainable Development (WSSD) (UN 2002) or the CBD Aichi Targets. All SDGs are applicable to the whole of the marine environment and to all states, whether developing or developed, island or continental, but their implementation must take into account states' national capacities, priorities and policies, and levels of development (UNGA 2015). The SDGs and their related targets are 'integrated and indivisible' and therefore must be considered and implemented as a whole (UNGA 2015). This means that the oceans, just like the other issues tackled by the 2030 Agenda, play a cross-cutting role across all SDGs, so that any SDG, including SDG 14, cannot be implemented in isolation from the other SDGs (Schmidt et al. 2017). Thus, SDG 14 provides a unique opportunity to consider, through the lens of ocean governance, the complex interlinkages between sustainability issues highlighted by the wide array of SDGs that are sometimes contradictory (Schmidt et al. 2017).

2.2. Ensuring Coherence across Global Processes for Marine Conservation

Although these processes differ in terms of their specific nature and scope, there exist numerous benefits for considering them holistically and coordinating efforts in achieving their objectives. Moreover, given their global scale and similarities, it is important to ensure coherence between actions (e.g., spatial coverage, sectoral coverage, and inclusion of key ocean governance principles) taken within these three processes to achieve ocean conservation. By considering these processes jointly and taking a coherent approach to their achievement, efficiency gains can also be made. These include utilizing data and information, and therefore resources, across multiple uses and functions, as well as building capacity to understand underpinning ocean science and implement and review the needed actions towards conservation goals.

The ongoing BBNJ negotiations represent a major opportunity for states to create a legally binding instrument by which to conserve and sustainably manage marine biodiversity in ABNJ, but also the potential for states to underpin actions taken to achieve global conservation goals. It may even be argued that the Aichi Targets and SDGs, particularly SDG 14, will not be fully achieved without the BBNJ Agreement, as ocean conservation is currently not fully delineated under the current legal framework. A critical difference is that the BBNJ Agreement will be a legally binding agreement, whereby the Aichi Targets and SDGs are non-binding. Through its legally binding nature, the BBNJ Agreement could: enhance the role of multi-level governance in ocean processes; ensure coordination and collaboration amongst states as well as relevant organizations; offer the means for new arrangements for data collection and information exchange; support capacity building and financing for ocean conservation and related initiatives (e.g., research vessels, data platforms, etc.); as well as ensure that key lessons and best practices are shared across states, organizations, and stakeholders. Ultimately, it may also be expected that conservation gains within the Aichi Targets or SDG 14 (e.g., the establishment of marine protected areas within states' national waters) will contribute to the objectives within the BBNJ process to protect marine biodiversity in ABNJ due to the ecological connectivity of the ocean, and vice versa. Furthermore, by taking such connections into account (e.g., when establishing marine protected areas), ecological links between national marine waters and ABNJ can be included, thus creating synergistic effects, such as conservation goals and restoration effects.

3. Improving Ocean Governance to Support Global Processes and Marine Conservation Goals

Achieving global conservation goals will require the international community to take a holistic approach to address sustainability issues. States and organizations will therefore need to go beyond established single-sector and state-centric ocean governance approaches (Wright et al. 2017). States cannot effectively manage ocean challenges working in isolation as marine ecosystems (e.g., the Sargassum ecosystem in the Atlantic, the Costa Rica Dome in the Pacific, etc.) and marine species (e.g., fish stocks, migratory species such as turtles, sharks or marine mammals) do not respect national borders, and threats to biodiversity are often transboundary in nature (e.g., marine pollution) (Boteler et al. 2019a). Thus, enhanced cooperation and coordination, particularly at the marine region level and across sectors, offer an opportunity for improving the conservation and sustainable use of marine biodiversity (Boteler et al. 2019b).

Regional and sectoral organizations can support the achievement of global conservation goals and targets by developing, implementing and enforcing regionally or sectoral-based agreements in alignment with global targets (Gjerde et al. 2018; Durussel et al. 2018). Such agreements could reflect the specificity of each region, their challenges and needs, and allow organizations to develop new initiatives to strengthen or complement existing efforts, and even adopt more stringent measures when needed (Gjerde et al. 2018; Durussel et al. 2018). Regional organizations have a long history of bringing states and regional bodies together to collaborate on marine issues, including through conducting scientific assessments, forming working groups, issuing protocols and guaranteeing compliance (Gjerde et al. 2018; Durussel et al. 2018). Such cooperation and coordination amongst actors can also increase transparency within decision-making processes. Thus, efforts at the marine regional level and through sectoral organizations can, and should, play a crucial role in global ocean governance and delivering ocean sustainability by providing for cooperation and coordination across organizations and across boundaries (UN Environment 2017). The current BBNJ negotiations offer a unique opportunity to build the institutional arrangements or mechanisms essential to creating a holistic approach to ocean governance and enable the achievement of Aichi Targets and targets of the Post-2020 Global Biodiversity Framework and SDGs.

The regional level can offer a particularly efficient means to implement global conservation goals. Ensuring the implementation of regionally agreed targets and indicators that are in line with globally agreed goals will be important to deliver the global conservation goals, while taking into account the priorities, challenges and needs of the regions (Boteler et al. 2019b; Institute for Advanced Sustainability Studies e.V. (IASS) et al. 2020). Implementation at the regional level is also particularly well-suited as it can build on existing regional initiatives and thereby ensure strengthened regional cooperation amongst stakeholders and across sectors (Boteler et al. 2019b; Institute for Advanced Sustainability Studies e.V. (IASS) et al. 2020). The regional level could also be used as a regional follow-up and review mechanism to monitor and track down the achievement of global conservation goals, including SDG 14 (UN Environment 2018; Unger et al. 2017).

3.1. Coordinating Efforts and Taking Joint Action

Ocean governance is complex and evolving, meaning that a diverse range of contexts, interests, and capacities must be coordinated (Wright et al. 2017). The costs to coordinate and cooperate across this complex governance system can be costly, both in human and financial resources, ultimately impeding the achievement of tangible benefits for ocean sustainability (Wright et al. 2017). Indeed, limited resources are a common problem for many organizations and their contracting parties, and developing needed capacities and ensuring long-term funding for strategic global, or national, processes is a challenge (Wright et al. 2017). Although cooperation and coordination of efforts can be expensive, working collaboratively can also create new value for organizations (e.g., access to new data, capacity building, sharing of best practices and resources).

Coordination arrangements could be created or improved to pursue ecosystem-based management in coastal waters and ABNJ (Gjerde and Wright 2019). Regional arrangements have been shown to build understanding and political support for ocean governance, provided they also build links with regional multi-purpose organizations (UN Environment 2017). Cross-sectoral coordination can foster dialogue and exchange amongst stakeholders, thereby helping to build trust and political will, and can lead to the development of joint programs of work and largescale planning projects (Gjerde et al. 2018). It is also necessary to foster collaboration at the national level amongst ministries so that states take a harmonized position in the various regional, sectoral and international organizations (Gjerde et al. 2018). This can be a major challenge, preventing a coherent approach to ocean governance and sustainable management. This underscores the need to strengthen capacity at the national level in an effort to ensure that national representatives can meaningfully participate in and contribute to regional, sectoral and global processes (Gjerde et al. 2018).

Furthermore, strengthening intra-regional, inter-regional and region-to-global cooperation will be crucial. Establishing dialogue platforms are an option to facilitate learning processes and to gather organizations and actors from different regions to broaden the scope of existing approaches and develop new solutions. Such an approach provides an opportunity for different actors to meet informally to share experiences and good practices, discuss common initiatives, highlight options to tackle key challenges, and identify pathways toward improved cooperation for the achievement of global conservation goals (Durussel et al. 2018).

3.2. Capacity Building and Information Exchange as a Cornerstone for Ocean Action

Capacity building is a cross-cutting topic throughout the 2030 Agenda and referenced in SDG 14 and many other SDGs, especially in SDG 17 (Cicin-Sain et al. 2018a). Capacity building is a long-term and continuous 'process by which individuals, organizations, institutions and societies develop abilities to perform functions, solve problems and set and achieve objectives' (UN Economic and Social Council 2006). As such, 'the development of a country's human, scientific, technological, organizational, institutional and resource capabilities' forms the basis for the implementation of global conservation goals (Cicin-Sain et al. 2018a). The transfer of marine technology is one of the tools that can be used to build capacities in countries where access to data and technology is limited (Cicin-Sain et al. 2018a). Through the negotiation of a future BBNJ Agreement under UNCLOS, states will have the opportunity to legally strengthen these issues by establishing more detailed provisions on capacity building and technology transfer than those that can currently be found in UNCLOS, including a set of requirements and measures to build capacity and ensure the transfer of marine technology in developing countries, including small island developing states (SIDS) and less developed countries (LDCs) (especially in regard to Art7, Art10, Art11, Art42, Art43, Art44, Art45, Art46, Art47, Art49, Art51, and Art52 in (UNGA 2019). These legally binding provisions can contribute to setting basic requirements for capacity building and technology transfer and help to meet the goals of the Aichi Targets (and targets of the Post-2020 Global Biodiversity Framework) and the SDGs.

The regional level, as well as sectoral organizations, can greatly contribute to implementing these provisions and ensuring that they adequately reflect the reality and needs of the regions or stakeholders (Institute for Advanced Sustainability Studies e.V. (IASS) et al. 2020). Regular capacity building workshops can underpin the ongoing exchange of knowledge and data. At the same time, initiatives are needed to strengthen national, regional and sectoral institutions, as well as individual capacity, to ensure that national representatives are able to effectively participate in sectoral, regional and global processes and to design and implement actions towards global objectives (Gjerde et al. 2018).

Increased support for scientific cooperation programs could improve the ability of national, regional and sectoral organizations, to implement ecosystem-based management approaches. The regional level, for instance, could underpin this by establishing, or expanding, regional scientific knowledge hubs, similar to the International Council for the Exploration of the Sea (ICES). Such initiatives could provide regionally targeted scientific and technical advice, and disseminate knowledge and data to different regional organizations, thereby boosting cross-sectoral cooperation and exchange (Gjerde et al. 2018).

3.3. Long-Term and Consistent Financing Is an Enabler for Action

Ensuring long-term and consistent funding for ocean measures, including for science and capacity building, that deliver the necessary protection of marine biodiversity and support common ocean conservation objectives, is an essential component and enabler for other ocean actions (Laffoley et al. 2019). Total funding available from public sources is insufficient to deliver the agreed marine protection goals. Innovative financing sources, including from capital markets, offer significant potential to support the delivery of ocean solutions across initiatives, including for coastal ecosystems in national waters, as well as for ABNJ (Thiele and Gerber 2017). Lessons can be drawn from sustainable development and climate financing approaches which are already in place. Potential sources include accessing private capital, as well as creating new mechanisms to inject funding into ocean initiatives. For example, climate bonds demonstrate how private sector finance for renewables has been used. Potential "blue bonds" for ocean solutions (Roth et al. 2019) can provide a means to provide capital to conservation projects, and could include performance-based components that would also allow the sharing of risk and encourage an efficient delivery of actions. The Nordic Investment Bank successfully raised US$ 200 million through a blue bond to deliver cheaper funding to multiple water treatment projects along the Baltic coast, and the Seychelles used a sovereign blue bond to help fund the implementation of marine protection. Such efforts could also bring together public and private actors in partnerships, which in turn, can support greater inclusion of stakeholders and transparency (Cicin-Sain et al. 2018b). The BBNJ process needs to include robust financing mechanisms, in order to develop new funding initiatives for ABNJ efforts. It needs to consider the Aichi Targets and SDGs and create links to enable, or enhance, the financing of ongoing and future initiatives underway through these processes (Claudet et al. 2019).

3.4. Lessons Learned from Past and Ongoing Marine Initiatives Should Be Leveraged for the Future

The analysis of ocean governance approaches and sharing of experience, in particular at the regional level or from sectoral organizations, can provide useful lessons that can facilitate the further development of new initiatives and help to strengthen existing frameworks (Mahon et al. 2015; Mahon and Fanning 2019). It also helps to inform the construction of efficient and effective means to support the joint achievement of objectives for ocean conservation, through the BBNJ Agreement, the Aichi Targets and targets of the Post-2020 Global Biodiversity Framework, and the SDGs. In many cases, lessons or options to overcome challenges may be regionally specific (e.g., due to available funding), while there is still a strong case for identifying common challenges and exchanging key lessons gained within such a specific context. In particular, lessons can be gained on effective arrangements for cooperation and coordination between organizations; achievements in successful capacity building efforts; the development of science and tools to inform decision making; the role played by champions and leaders with the political will to drive processes and gather support for improved management; developing innovative financing mechanisms; and the importance of developing a dynamic science–policy interface that can provide policy-relevant scientific information to decision makers and stakeholders. Such lessons need to be harvested from established processes and institutions and organizations—potentially through organizing workshops and events for dialogue and exchange or through funding research and development projects by which to collect, assess, and disseminate key lessons or best practices.

4. Conclusions

Strengthened collaboration and cooperation between global, regional and sectoral organizations will be necessary to enhance ocean governance and to underpin actions for the conservation and sustainable use of BBNJ, the Aichi Targets and targets of the Post-2020 Global Biodiversity Framework, and the SDG Targets. The facilitation of joint actions and coordinated efforts through dialogue platforms and participatory learning processes to share experiences and good practices will also be crucial for the achievement of global conservation goals. Capacity building and information exchange, including through the transfer of technology, expanded support for scientific cooperation programs or regional scientific knowledge hubs, and long-term and consistent funding for ocean initiatives, can further help to boost cross-sectoral and multi-level cooperation and exchange, which represent important cornerstones for ocean action. The current BBNJ negotiations provide the opportunity to create institutional arrangements for cross-sectoral collaboration embedded in a binding legal instrument. Such a collaborative approach could help to overcome the currently fragmented approach to ocean governance and thereby foster critical conditions to achieve the Aichi Targets and targets of the Post-2020 Global Biodiversity Framework, and SDGs.

Regional and sectoral organizations can help to underpin global conservation goals and targets by developing, implementing and enforcing regional or sectoral-based agreements. Enhanced cooperation at the scale of marine regions can play a particular role in specifying global ambitions and objectives into relevant and regionally achievable, harmonized and measurable targets. Ensuring the implementation of regionally agreed targets and indicators will be important to deliver the global conservation goals while taking into account the priorities, ecological characteristics, challenges and needs of the regions. Regional ocean governance strategies or cooperation platforms should be established in support of the 2030 Agenda and to bring together states, regional and global organizations, different sectors, and a broad spectrum of stakeholders, including non-governmental organizations, research centers, and private sector actors, and donors. A follow-up and review mechanism at the regional level can also be relevant to monitor and track down the achievement of global conservation goals and their legal regional implementation.

The BBNJ negotiations represent a major opportunity for states to create a legally binding instrument that can help to underpin actions taken to achieve global conservation goals. It can particularly outline more detailed provisions on capacity building, technology transfer and funding initiatives that are currently found in UNCLOS. Without this agreement, it can be argued that the Aichi Targets and targets of the Post-2020 Global Biodiversity Framework and SDGs will not be fully achieved as ocean conservation is currently not fully delineated under the current legal framework, in particular in ABNJ which covers almost half of the Earth's surface.

Author Contributions: Conceptualization for this chapter was done by B.B., C.D. and S.U.; Writing—Original Draft Preparation was done by B.B., C.D., S.U., and T.T.; Writing—Review and Editing was done by B.B., C.D. and S.U.

Funding: This research builds on ongoing work undertaken as part of the Ocean Governance research group at the Institute for Advanced Sustainability Studies (IASS), which is funded by the German Federal Ministry for Education and Research (BMBF) through its Research for Sustainable Development (FONA) program, as well as the State of Brandenburg Ministry for Science, Research and Culture. The Marine Regions Forum 2019 was also an important platform contributing to many of the thoughts and ideas reflected in this chapter and was sponsored by the German Federal Ministry for Environment, Nature Conservation and Nuclear Safety (BMU), the German Environment Agency (UBA), the European Maritime and Fisheries Fund (EMFF) and the German Ministry for Education and Research (BMBF) through its Research for Sustainable Development (FONA) program, as well as the State of Brandenburg Ministry for Science, Research and Culture. This work also builds on research undertaken in the STRONG High Seas project which is funded by the German Federal Ministry for the Environment, Nature Conservation and Nuclear Safety (BMU) through the International Climate Initiative. This work contributes to the Partnership for Regional Ocean Governance (PROG), a partnership hosted by UN Environment, the Institute for Advanced Sustainability Studies (IASS), the Institute for Sustainable Development and International Relations (IDDRI), and TMG—Think Tank for Sustainability.

Conflicts of Interest: The authors declare no conflict of interest. The founding sponsors had no role in the design of the study; in the collection, analysis, or interpretation of data; in the writing of the manuscript; or in the decision to publish the results.

References

Boteler, Ben, Ross Wanless, Maria Dias, Tim Packeiser, Adnan Awad, Beatriz Yannicelli, Padilla Zapata, Alfonso Luis, Jaime Aburto, Isabelle Seeger, and et al. 2019a. *'Ecological Baselines for the Southeast Atlantic and Southeast Pacific: Status of Marine Biodiversity and Anthropogenic Pressures in Areas Beyond National Jurisdiction'*. STRONG High Seas Project. Potsdam: IASS.

Boteler, Ben, Joseph Appiott, Carole Durussel, Takehiro Nakamura, and Sebastian Unger. 2019b. Regional Ocean Governance in the Post-2020 Biodiversity Framework. Paper presented at the 2020 Ocean Pathways Week, Montreal, QC, Canada, November 11–15.

CBD/COP/DEC/14/34. 2018. Decision Adopted by the Conference of the Parties to the Convention on Biological Diversity: Comprehensive and Participatory Process for the Preparation of the Post-2020 Global Biodiversity Framework. Available online: https://www.cbd.int/doc/decisions/cop-14/cop-14-dec-34-en.pdf (accessed on 12 November 2019).

Cicin-Sain, Biliana, Marjo Vierros, Miriam Balgos, Alexis Maxwell, Meredith Kurz, Tina Farmer, Atsushi Sunami, Miko Maekawa, Iwao Fujii, Awni Benham, and et al. 2018a. Policy Brief on Capacity Development as a Key Aspect of a New International Agreement on Marine Biodiversity Beyond National Jurisdiction (BBNJ). December. Available online: http://www.fao.org/fileadmin/user_upload/common_oceans/docs/policy-brief-on-bbnj-capacity-development-aug-2018.pdf (accessed on 12 November 2019).

Cicin-Sain, Biliana, Alexis Maxwell, Miriam Balgos, Meredith Kurz, Brian Cortes, Vanessa Knecht, Carol Turley, Tarub Bahri, Dorothee Herr, Kirsten Isensee, and et al. 2018b. *Assessing Progress on Ocean and Climate Action: 2018. A Report of the Roadmap to Oceans and Climate Action (ROCA) Initiative*. Available online: https://rocainitiative.files.wordpress.com/2018/12/roca-progress-report-2018.pdf (accessed on 12 November 2019).

Claudet, Joachim, Laurent Bopp, William Cheung, Rodolphe Devillers, Elva Escobar-Briones, Peter Haugan, Johanna Heymans, Valérie Masson-Delmotte, Nele Matz-Lück, Patricia Miloslavich, and et al. 2019. A Roadmap for Using the UN Decade of Ocean Science for Sustainable Development in Support of Science, Policy, and Action. *One Earth* 2: 34–42. [CrossRef]

Di Franco, Antoinio, Bronwyn Gillanders, Giuseppe De Benedetto, Antonio Pennetta, Guila A De Leo, and Paolo Guidetti. 2012. Dispersal Patterns of Coastal Fish: Implications for Designing Networks of Marine Protected Areas. *PLoS ONE* 7: e31681. [CrossRef] [PubMed]

Dunn, Daniel, Autumn-Lynn Harrison, Corrie Curtice, Sarah DeLand, Ben Donnelly, Ei Fujioka, Eleanor Heywood, Connie Kot, Sarah Poulin, Meredith Whitten, and et al. 2019. The importance of migratory connectivity for global ocean policy. *Proceedings of the Royal Society B* 286: 20191472. [CrossRef] [PubMed]

Durussel, Carole, Glen Wright, Nicole Wienrich, Ben Boteler, Sebastian Unger, and Rochette Julien. 2018. *Strengthening Regional Governance for the High Seas: Case Studies of Opportunities and Challenges for the Southeast Atlantic and Southeast Pacific.* STRONG High Seas Project. Potsdam: IASS.

Gjerde, Kristina, and Glen Wright. 2019. *Towards Ecosystem-based Management of the Global Ocean Strengthening Regional Cooperation through a New Agreement for the Conservation and Sustainable Use of Marine Biodiversity in Areas Beyond National Jurisdiction.* STRONG High Seas Project. Potsdam: IASS.

Gjerde, Kristina, Ben Boteler, Carole Durussel, Julien Rochette, Sebastian Unger, and Glen Wright. 2018. *Conservation and Sustainable Use of Marine Biodiversity in Areas Beyond National Jurisdiction: Options for Underpinning a Strong Global BBNJ Agreement through Regional and Sectoral Governance.* STRONG High Seas Project. Potsdam: IASS.

HELCOM. 2016. Ecological coherence assessment of the Marine Protected Area network in the Baltic. In *Baltic Sea Environment Proceedings.* No. 148. Helsinki: Helsinki Commission.

Institute for Advanced Sustainability Studies e.V. (IASS), Institute for Sustainable Development and International Relations (IDDRI), and TMG—ThinkTank for Sustainability (TMG). 2020. *Marine Regions Forum 2019: Achieving a Healthy Ocean—Regional Ocean Governance Beyond 2020.* Conference Report—Marine Regions Forum 2019. Berlin: Institute for Advanced Sustainability Studies e.V. (IASS), September 30—October 2. [CrossRef]

IPBES. 2019. *Summary for Policymakers of the Global Assessment Report on Biodiversity and Ecosystem Services of the Intergovernmental Science-Policy Platform on Biodiversity and Ecosystem Services.* Edited by Sandra Díaz, Josef Settele, Eduardo Brondizio, Hien Ngo, John Agard, Almut Arneth, Patricia Balvanera, Kate Brauman, Stuart Butchart, Kai Chan and et al. IPBES Secretariat. Bonn: IPBES, p. 56.

IPCC. 2019. *IPCC Special Report on the Ocean and Cryosphere in a Changing Climate.* Edited by Hans-Otto Pörtner, Debra Roberts, Valérie Masson-Delmotte, Panmao Zhai, Melinda Tignor, Elvira E. Poloczanska, Katja Mintenbeck, Andrés Nicolai Alegría, Okem Maike, Petzold Andrew and et al. Nora Geneva: IPCC, in press.

Laffoley, Dan, John Baxter, Diva Amon, Duncan Currie, Craig Downs, Jason Hall-Spencer, Harriet Harden-Davies, Richard Page, Chris Reid, Callum Roberts, and et al. 2019. Eight urgent fundamental steps to recover ocean sustainability, and the consequences for humanity and the planet of inaction or delay. *Aquatic Conservation* 30: 194–208. [CrossRef]

Mahon, Robin, and Lucia Fanning. 2019. Regional ocean governance: Integrating and coordinating mechanisms for polycentric systems. *Marine Policy* 107: 103589. [CrossRef]

Mahon, Robin, Lucia Fanning, Kristina Gjerde, Oran Young, Michael Reid, and Selicia Douglas. 2015. *Transboundary Waters Assessment Programme (TWAP) Assessment of Governance Arrangements for the Ocean, Volume 2: Areas Beyond National Jurisdiction*. IOC Technical Series; Paris: UNESCO-IOC, vol. 119, p. 91.

Neumann, Barbara, and Sebastian Unger. 2019. From voluntary commitments to ocean sustainability. *Science* 363: 35–6. [CrossRef] [PubMed]

Popova, Ekaterina, David Vousden, Warwick Sauer, Essam Mohammed, Valerie Allain, Nicola Downey-Breedt, and Andrew Yool. 2019. Ecological connectivity between the areas beyond national jurisdiction and coastal waters: Safeguarding interests of coastal communities in developing countries. *Marine Policy* 104: 90–102. [CrossRef]

Rogers, Troy, Anthony Fowler, Michael Steer, and Bronwyn Gilanders. 2019. Spatial connectivity during the early life history of a temperate marine fish inferred from otolith microstructure and geochemistry. *Estuarine, Coastal and Shelf Science* 227. [CrossRef]

Roth, Natalie, Torsten Thiele, and Martin Von Unger. 2019. *Blue Bonds: Financing Resilience of Coastal Ecosystems: Key Points for Enhancing Finance Action*. Brussels: IUCN.

Schmidt, Stefanie, Barbara Neumann, Yvone Waweru, Carole Durussel, Sebastian Unger, and Martin Visbeck. 2017. SDG 14—Conserve and Sustainably Use the Oceans, Seas and Marine Resources for Sustainable Development. In *A Guide to SDG Interactions: From Science to Implementation*. Edited by David Griggs, Mâns Nilsson, Anne-Sophie Stevance and David McCollum. Paris: International Council for Science.

Thiele, Torsten, and Leah Gerber. 2017. Innovative Financing for the High Seas. *Aquatic Conservation* 27: 89–99. [CrossRef]

UN. 2002. *United Nations Report of the World Summit on Sustainable Development. Johannesburg, South Africa. A/CONF.199/20**. New York: UN, August 26—September 4.

UN Economic and Social Council. 2006. *Definition of Basic Concepts and Terminologies in Governance and Public Administration, E/C.16/2006/4, 5th Session, Agenda Item 5*. New York: UN Economic and Social Council, January 5.

UN Environment. 2017. *Regional Seas Programmes Covering Areas beyond National Jurisdiction*. Regional Seas Reports and Studies No. 202. Nairobi: UN Environment.

UN Environment. 2018. *Regional Seas Follow up and Review of the Sustainable Development Goals (SDGS)*. UN Environment Regional Seas Reports and Studies No. 208. Nairobi: UN Environment.

UNEP/CBD/COP/DEC/X/2. 2010. Decision Adopted by the Conference of the Parties to the Convention on Biological Diversity at its Tenth Meeting. X/2. The Strategic Plan for Biodiversity 2011–2020 and the Aichi Biodiversity Targets. Available online: https://www.cbd.int/doc/decisions/cop-10/cop-10-dec-02-en.pdf (accessed on 12 November 2019).

UNGA. 1992. *United Nations General Assembly, Report of the United Nations Conference on Environment and Development, Conference on Environment and Development, A/CONF.151/26 (Vol. II) (13 August 1992) Chapter 17 ('Protection of the Oceans, All Kinds of Seas, Including Enclosed and Semi-Enclosed Seas, and Coastal Areas and the Protection, Rational Use and Development of their Living Resources').* New York: UNGA.

UNGA. 2015. *United Nations General Assembly, Transforming Our World: The 2030 Agenda for Sustainable Development, Resolution adopted by the General Assembly on 25 September 2015, GA Res 70/1, 70th session, Agenda Items 15 and 116, A/Res/70/1 (21 October 2015).* New York: UNGA.

UNGA. 2019. *Revised Draft Text of an Agreement under the United Nations Convention on the Law of the Sea on the Conservation and Sustainable Use of Marine Biological Diversity of Areas Beyond National Jurisdiction, A/CONF.232/2020/3 (18 November 2019).* New York: UNGA.

UNGA/RES/69/292. 2015. UNGA/RES/69/292 on the Development of an International Legally-Binding Instrument under the United Nations Convention on the Law of the Sea on the Conservation and Sustainable Use of Marine Biological Diversity of Areas Beyond National Jurisdiction. Available online: https://undocs.org/en/a/res/69/292 (accessed on 12 November 2019).

UNGA/RES/72/249. 2018. UNGA/RES/72/249 on an International legally Binding Instrument under the United Nations Convention on the Law of the Sea on the Conservation and Sustainable Use of Marine Biological Diversity of Areas Beyond National Jurisdiction. Available online: https://undocs.org/en/a/res/72/249 (accessed on 12 November 2019).

Unger, Sebastian, Alexander Müller, Julien Rochette, Stefanie Schmidt, Jana Shackeroff, and Glen Wright. 2017. *Achieving the Sustainable Development Goal for the Oceans.* IASS Policy Brief 1/2017. Potsdam: IASS February.

Wright, Glen, Stefanie Schmidt, Julien Rochette, Jana Shackeroff, Sebastian Unger, Yvonne Waweru, and Alexander Müller. 2017. *Partnering for a Sustainable Ocean: The Role of Regional Ocean Governance in Implementing SDG14.* (Summary for Decision Makers) IDDRI, IASS, TMG & UN Environment. Potsdam: PROG.

Climate Change and Its Impact on the Ocean

Martin Visbeck and Sigrid Keiser

1. Introduction

Over the last century, the world's human population has grown rapidly, and along with affluent lifestyles, the demand for energy has been growing. Since most of this energy is obtained from various kinds of fossil fuels, the atmospheric concentration of CO_2 has been rapidly increasing also. On land, the growing population has led to the significant loss of the natural landscape in favor of settlements and land culture. This development has resulted in many dimensions of pressure on our environment (e.g., Rockström et al. 2009). Climate change, loss of biodiversity and ocean acidification, as well as land-based pollution, all affect the ocean. The rapid depletion of natural resources and environmental pollution have led the global community to sound the alarm bells and articulate a number of global agreements to protect the planet and the life-supporting ecosystem services it provided to humanity. The Paris Climate Accord, the Sendai Framework for Disaster Risk Reduction and the all-encompassing 2030 Agenda for Sustainable Development with its Sustainable Development Goals (SDGs) all set out ambitious goals for the people and planet. The recognition that the ocean space also plays an important role for humanity in general, and its particular central role for small island developing states, has led to the inclusion of an explicit SDG for the ocean (Visbeck et al. 2014). However, the ocean dimension of the 2030 Agenda goes beyond SDG 14, and there are significant linkages across the SDGs. Some of the goals reinforce each other, while others are in conflict (Nilsson et al. 2016). Schmidt et al. (2017) provide a deeper analysis and, amongst others, highlight the connection between SDG 14 and SDG 13 as a tight nexus between ocean and climate.

Geographically, a typical world map in the often-used Mercator projection seems to suggest that there could be five separated oceans. However, they are in fact well-connected ocean basins and part of a single global ocean system (Figure 1). Ocean circulation connects all ocean basins and regions, and therefore, the ocean is host to the largest connected ecosystem of our planet and covers more than two-thirds of the Earth's surface. This large surface area combined with an enormous volume of water and mass makes the ocean a very important part of the global climate system (e.g., Schmitt 2018).

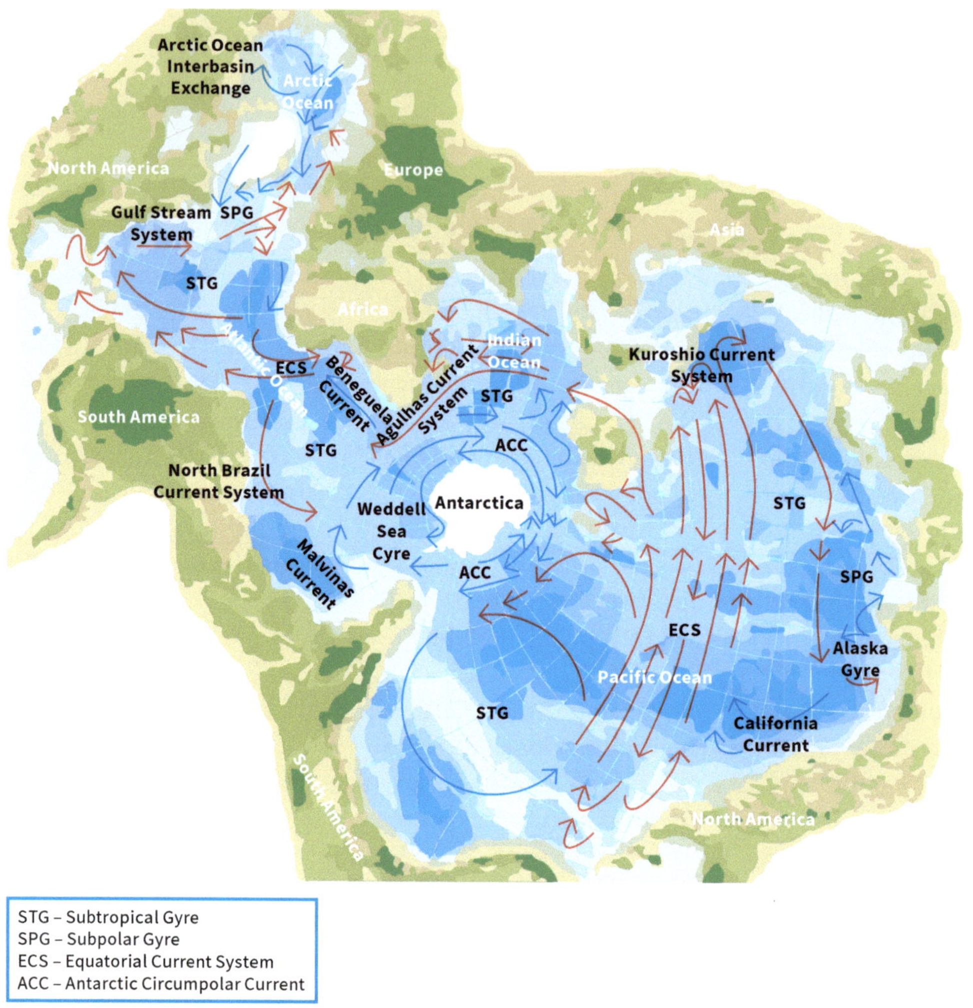

Figure 1. Horizontal ocean currents: the ocean is traversed by a large-scale overturning circulation that sets the rate at which the deep ocean interacts with the atmosphere and is, therefore, crucial for the climate system Source: Figure by Jamie Oliver, British Antarctic Survey, 2020; used with permission.

The rising emissions of carbon dioxide (CO_2), methane and other greenhouse gases have led to planetary warming. These greenhouse gases are preventing some of the heat radiated from the Earth's surface from escaping into space, heating the atmosphere and increasing the back radiation onto the Earth's surface. Recent estimates of this change in planetary energy changes have revealed that more than 93% of warming is found in the upper and deep ocean. The large volume and high heat capacity of water make the ocean the largest sink of this extra warming. In addition to buffering the heat, the ocean also directly stores almost 30% of the human emitted carbon dioxide. Taken together, these two effects have significantly slowed down human-induced climate change in the atmosphere and on land. At the same time, this warming changes the physics, biochemistry and marine ecosystems with an already noticeable impact on the marine ecosystem services.

This fundamental role of the ocean in climate variability and change has prompted member states to ask for a topical report of the Intergovernmental Panel on Climate Change called the "Special Report on the Ocean and Cryosphere in a Changing Climate" (IPCC 2019). We will make extensive use of this detailed report to highlight some of the specifics throughout this chapter.

More generically, the ocean moderates the seasonal cycle of our climate and is responsible for what we call a maritime climate with warmer winters and cooler summers. It enables the presence of monsoon systems and moderates global rainfall and regional weather variability. It plays a key role in coupled ocean–atmosphere phenomena, such as El-Niño, and enables their predictability. On longer timescales, the ocean is a critical pacemaker from decade long megadroughts, century long cold spells through the cycles of ice ages and beyond.

The ocean also interacts with other parts of the climate system—land, atmosphere, sea ice and the marine ecosystem. Its global circulation system connects ocean basins (Figure 1) and the upper ocean with the deeper waters (Figure 2). Ocean currents transport heat from warm to cold regions, and thus influence the release of heat and moisture to the atmosphere. They are connected with the atmosphere and modulate the wind system that in turn drives ocean currents. We speak of a coupled ocean–atmosphere climate system. In addition, higher levels of atmospheric CO_2 directly lead to increasing levels of dissolved CO_2 in the upper ocean, leading to a change in the ocean chemistry that lowers the pH, a process known as ocean acidification. This ocean acidification describes a movement of pH from a slightly basic pH of the sea water (pH about 8 and >7) towards pH-neutral conditions rather than turning acidic (pH < 7).

The effects of a changing ocean have a high impact on our lives. Humans depend on the ocean directly by living at the sea and indirectly through profiting from the ocean ecosystem services (e.g., Visbeck et al. 2014). Any change in the ocean will also affect those services, sometimes directly and sometimes more indirectly, through complex interactions with the climate system. For many marine ecosystems, the combined stress from climate change, overuse, habitat destruction and pollution lead to dramatic shifts in ways that are often not understood. From a precautionary principles perspective, climate change needs to be minimized by decisive mitigation action in order to secure ocean ecosystem services for future generations. From the perspective that the human effects on the ocean and use of the ocean services need to become more sustainable, clear guiding principles can be articulated to inform sustainable development for humanity.

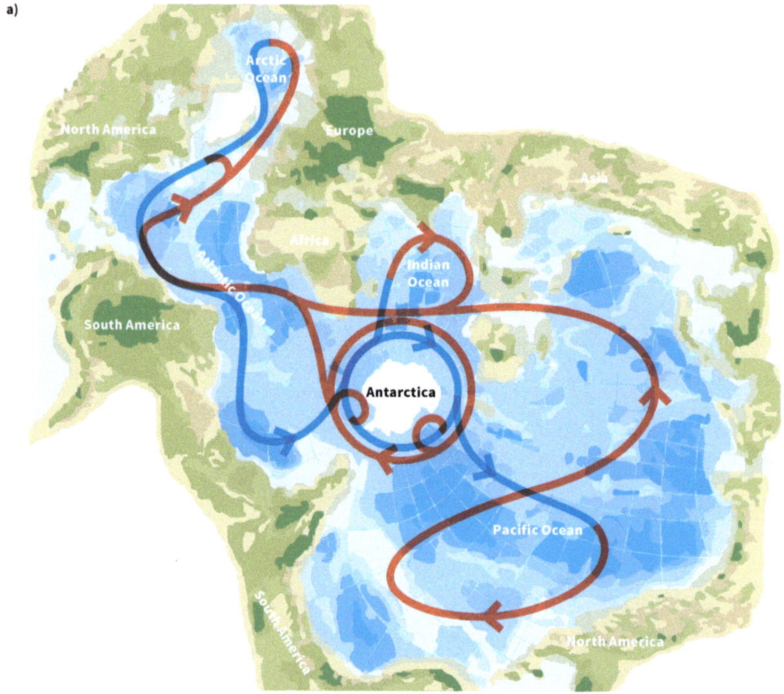

Figure 2. *Cont.*

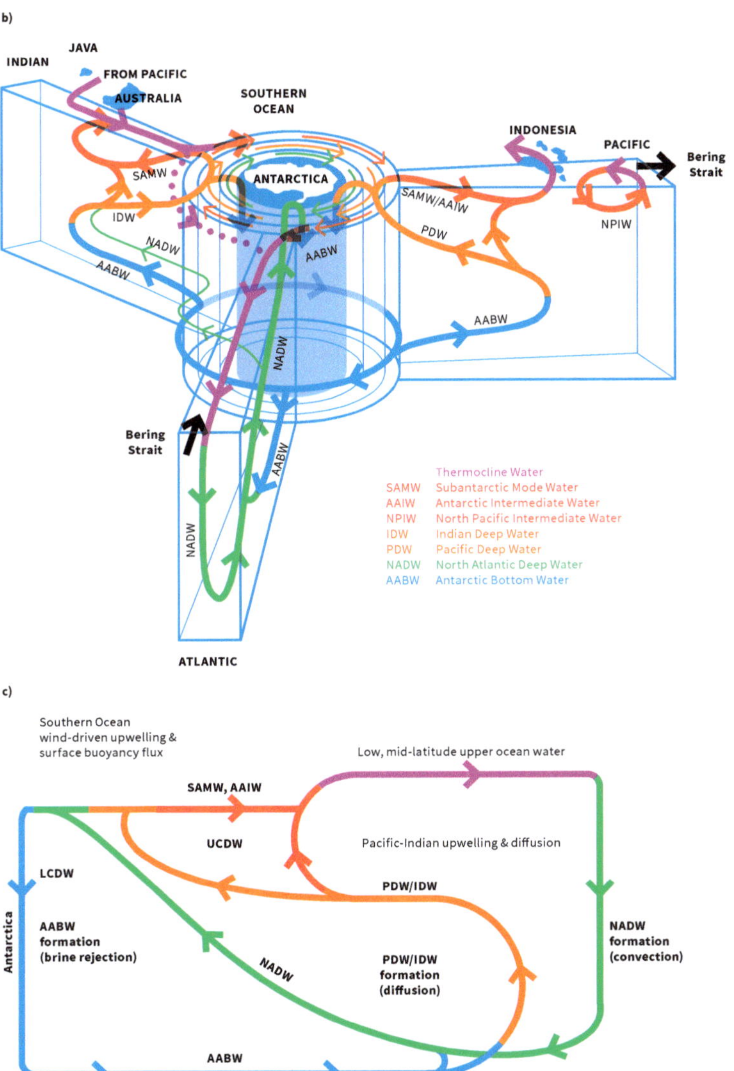

Figure 2. (a) Meridional ocean circulation: the ocean is traversed by a large-scale overturning circulation. This circulation sets the rate at which the deep ocean interacts with the atmosphere and is, therefore, essential for the climate system. The red arrows

depict a simplified pathway of the warmer water close to the surface, and the blue arrows show the spreading of cold water at depths. (**b**) Sketch of zonally averaged vertical circulation for three ocean basins shows the interconnection between the different water masses of the global ocean. (**c**) Replotting of the Atlantic Basin. Surface waters: purple; intermediate waters: red; NADW: green; Indian Ocean Deep Water (IDW): orange; Pacific Deep Water (PDW): orange; and Antarctic Bottom Water (AABW): blue Source: (**a**) adapted from (. Meredith 2019); figure by Jamie Oliver, British Antarctic Survey, 2020; used with permission; (**b,c**) figure by (Talley 2013); used with permission.

Knowledge about anthropogenic-induced climate change has been assessed by the Intergovernmental Panel on Climate Change since 1990 in a series of extensive reports. The latest IPCC Synthesis Report was published in 2014. The overall observations reported in the report confirm that the atmosphere and the ocean have warmed, the amounts of snow and ice have diminished and sea level has risen (IPCC 2014). The recent Special Report of the IPCC with a focus on the ocean and cryosphere (Special Report on the Ocean and Cryosphere in a Changing Climate, IPCC 2019) also emphasizes that climate change has led, for instance, to increasing ocean heat content, sea level rise, ocean heatwaves, coral bleaching and melting of ocean-terminating glaciers and ice sheets around Greenland and Antarctica. More indirect but measurable impacts are growing oxygen minimum zones, and there is an expectation that the global ocean overturning circulation will slow down in the future. In the following, we will repeat and highlight some of the key findings from that report that are particularly relevant in the context of sustainability.

2. The Ocean as a Heat and CO_2 Buffer

Since the pre-industrial era, human activities such as burning fossil fuels, e.g., coal and oil; deforestation; and cement production have led to an increase in atmospheric concentrations of carbon dioxide (CO_2), methane and nitrous oxide. The IPCC Synthesis report states that their effects, "together with those of other anthropogenic drivers, have been detected throughout the climate system and are extremely likely to have been the dominant cause of the observed warming since the mid-20th century" (IPCC 2014). The IPCC 2018 states that approximately 1.0 °C of global warming above pre-industrial levels, with a likely range of 0.8 to 1.2 °C, has been caused by human activities and estimates that global warming is likely to reach 1.5 °C between 2030 and 2052 if it continues to increase at the current rate (IPCC 2018).

The ocean's role in global warming is associated with its heat capacity, heat transport in the ocean and the global water cycle. The ocean has absorbed more than 90% of the increased heat (e.g., IPCC 2019). Consequently, the atmospheric warming would be much more dramatic today without the ocean's uptake of heat. About two thirds of the excess heat have been absorbed within the upper 700 m of the ocean. However, the warming has also reached the deep ocean through deep-water formation regions and by changing the ocean circulation (IPCC 2019).

In the past 100 years, the ocean has absorbed a third of man-made carbon dioxide. Without the ocean as a sink, the ever-increasing CO_2 emissions would already have caused much more pronounced global warming (e.g., IPCC 2014).

In addition to the heat buffer effect, the ocean has absorbed one third of global CO_2 emissions (IPCC 2014). When the concentration of CO_2 in the atmosphere rises, the atmosphere ocean pCO_2 gradient increases, leading to increased ocean CO_2 uptake. However, this exchange process is also moderated by the absolute ocean temperature. The colder the seawater is, the more CO_2 can be dissolved in the ocean. This means, in reverse, that a warming ocean will decrease the CO_2 uptake potential of the ocean. Moreover, a warmer surface ocean will increase the ocean's stratification and will make it harder to tap into the deeper ocean layers that provide enormous CO_2 uptake potential. A significant part of the drawdown of surface CO_2-rich waters is facilitated by the overturning ocean circulation, which itself is expected to decrease. In summary, there is an expectation that the ocean might not be able to continue to uptake its current share of the anthropogenic CO_2 emissions, which would increase the temperature effect per emitted ton of CO_2 more strongly (IPCC 2019).

3. Ocean Currents Regulate Global Climate

The Arctic, Atlantic, Indian, Pacific and Southern Oceans are all interconnected and part of a single Global Ocean. The circulation of the ocean together with exchanges between the ocean and the atmosphere, and to a lesser degree, with the coastal systems, determine the distribution of heat, freshwater, nutrients, oxygen, carbon dioxide and dissolved chemical components around the planet. The large-scale circulation of the ocean comprises an interaction of the dominantly wind-driven upper-ocean circulation of the various gyres (Figure 1). The vertical exchange between the surface and deeper layers is more complex. A prominent example is the global meridional overturning circulation (MOC) connecting the sinking regions of the higher latitudes with the upwelling regimes around the globe. This interaction leads to a complex three-dimensional circulation throughout the global ocean.

For example, in the Atlantic sector, the MOC brings warm tropical water northward in both hemispheres, mitigating the temperature difference between the equator and pole in the North Atlantic while amplifying the pole-to-equator temperature difference in the South Atlantic. This transport of heat in the upper limb of the Atlantic MOC contributes to the Gulf Stream (indicated in Figure 1) and heating of Western Europe, in particular, the Nordic Countries and Seas (Figure 2).

Global warming can affect this circulation in two ways. First, the melting of glacial ice produces low salinity waters that, together with warmer temperatures, increase the ocean's stratification and reduce the formation of deep-water. Second, a warmer upper ocean globally makes it more difficult to mix and upwell the cold waters from below towards the surface. Thus, experts are expecting a slow-down of the MOC in the coming decades. However, the full impact of these changes in the physical ocean is not fully understood, and future ocean observation and better understanding are needed globally.

A changing climate will also result in an atmospheric wind system. This in turn will directly influence the upper ocean circulation. However, the large interannual variability of the wind systems to date is significantly larger than climate change-induced signals. There is some evidence for a poleward shift of all major boundary currents over the last century and some suggestions of an intensification of the trade winds. All of this will affect the connectivity of the open ocean with the coasts of its ecosystems with significant regional differences (IPCC 2014 synthesis report).

4. Climate Change Induces Challenges for the Future Ocean

Unabated carbon emissions from human activities causing ocean warming, ocean acidification and oxygen loss, with some evidence of changes in nutrient cycling and primary production, will lead to more climate change. Some of the key parameters with their observed trend are listed in Figure 3. Two scenarios, a high emission and low emission one, are compared in the time series in Figure 4. The different projections for the climate future are described by Representative Conservation Pathways (RCPs), all of which are considered possible depending on the volume of greenhouse gases emitted in the years to come. Only the lowest projection represents mitigation pathways compatible with the 1.5 °C warming limit of the Paris Agreement. In particular, the high emission scenarios will cause several challenges for the future of the ocean.

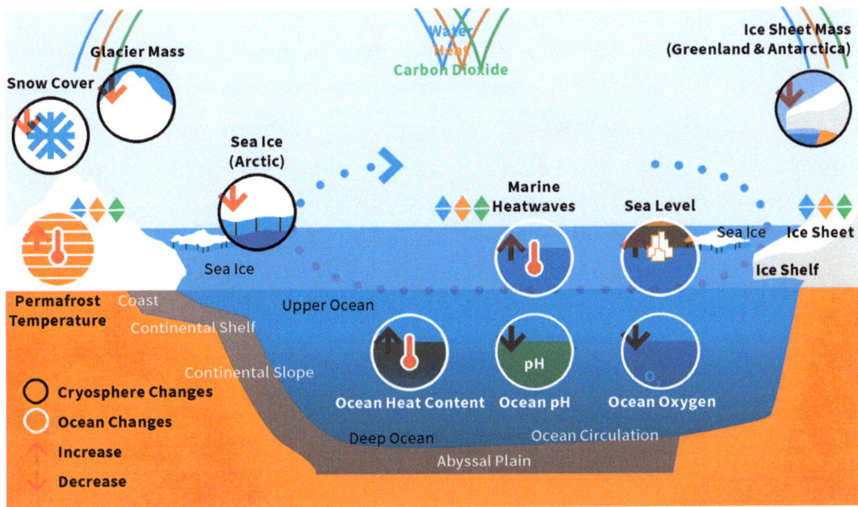

Figure 3. "Schematic illustration of key components and changes of the ocean and cryosphere, and their linkages in the Earth system through the global exchange of heat, water, and carbon. Climate change-related effects (increase/decrease indicated by arrows in pictograms) in the ocean include sea level rise, increasing ocean heat content and marine heatwaves, increasing ocean oxygen loss and ocean acidification. For illustration purposes, a few examples of where humans directly interact with ocean and cryosphere are shown" Source: Reprinted from (IPCC 2019, p. 43); used with permission. Figure TS.2 from IPCC 2019 Technical Summary.

- Ocean warming

The ocean has warmed progressively in recent decades. This trend is readily detectable in oceanic observations. The processes of global warming are scientifically well understood, and projections through climate models consistently show the global warming trend (IPCC 2019). The IPCC states: "These trends in the global average ocean temperature will continue for centuries after the anthropogenic forcing is stabilized" (IPCC 2019). The IPCC concludes that "this temperature increase corresponds to an uptake of over 90% of the excess heat accumulated in the Earth system over this period by the ocean and also causes it to expand and has contributed about 43% of the observed global mean sea level rise" (IPCC 2019). The IPCC summarizes that

The ocean has warmed unabated since 2005, continuing the clear multidecadal ocean warming trends. The warming trend is further confirmed by the improved ocean temperature measurements over the last decade. (…) By 2100 the ocean is very likely to warm by 2 to 4 times as much for low emissions and 5 to 7 times as much for the high emissions scenario compared with the observed changes since 1970. (…) The overall warming of the ocean will continue this century even after radiative forcing and stabilized mean surface temperatures. (IPCC 2019)

- Sea level rise

Global mean sea level is rising, with acceleration in recent decades (IPCC 2019). The IPCC report states that the sum of glacier and ice sheet contributions is now the dominant source of sea level rise followed by ocean warming (IPCC 2019). Future sea level rise caused by thermal expansion, melting of glaciers and ice sheets and land water storage changes is strongly dependent on which emission scenario is followed by society (IPCC 2019). Under all scenarios of the IPCC's projections—including those compatible with achieving the long-term temperature goal set out in the Paris Agreement—sea level rise will be faster at the end of the century (IPCC 2019). Projections of the IPCC conclude that the global mean sea level will rise between 0.29 and 0.59 m for low emissions and 0.61–1.10 m for high emissions by 2100 (IPCC 2019). However, sea level does not rise globally uniformly and varies regionally as "thermal expansion, ocean dynamics and land ice loss contributions will generate regional departures of about ±30% around the mean" (IPCC 2019). The differences from the global mean can be even greater in areas of rapid vertical land movements, including those caused by local anthropogenic factors, such as groundwater extraction (IPCC 2019). Therefore, the IPCC concludes that regional sea level rise is, in particular, a high risk to low-lying islands, coasts, cities and settlements, and needs response options and pathways to resilience and sustainable development along the coast (IPCC 2019).

- Ocean acidification

In the past 50 years, the ocean has taken up 20–30% of total carbon dioxide released into the atmosphere by human activities (IPCC 2019). However, as consequence, the average pH at the ocean surface has lowered from 8.2 to 8.1, which translates into a 30% increase in acidity (IPCC 2018).

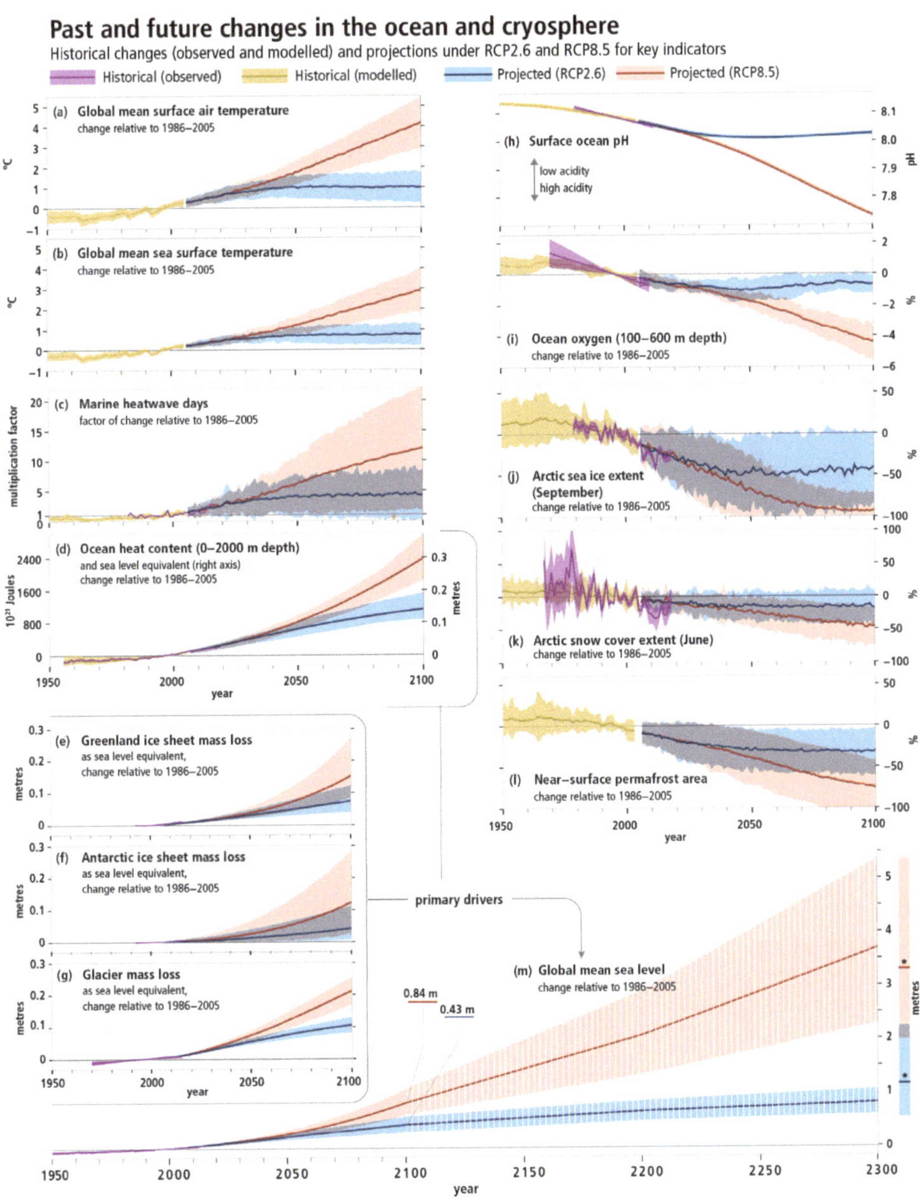

Figure 4. "Observed and modeled historical changes in the ocean and cryosphere since 1950, and projected future changes under low (Representative Concentration Pathways (RCP) 2.6; requires that carbon dioxide emissions start to decline by 2020

97

and go to zero by 2100) and high (RCP8.5) greenhouse gas emissions scenarios (emissions continue to rise throughout the 21st century). Changes are shown for: (**a**) Global mean surface air temperature change with likely range. Ocean-related changes with very likely ranges for (**b**) Global mean sea surface temperature change; (**c**) Change factor in surface ocean marine heatwave days; (**d**) Global ocean heat content change (0–2000 m depth). An approximate steric sea level equivalent is shown with the right axis by multiplying the ocean heat content by the global-mean thermal expansion coefficient ($\varepsilon \approx 0.125$ m per 1024 Joules) for observed warming since 1970; (**h**) Global mean surface pH (on the total scale). Assessed observational trends are compiled from open ocean time series sites longer than 15 years; and (**i**) Global mean ocean oxygen change (100–600 m depth). Assessed observational trends span 1970–2010 centered on 1996. Sea level changes with likely ranges for (**m**) Global mean sea level change. Hashed shading reflects low confidence in sea level projections beyond 2100 and bars at 2300 reflect expert elicitation on the range of possible sea level change; and components from (**e,f**) Greenland and Antarctic ice sheet mass loss; and (**g**) Glacier mass loss. Further cryosphere-related changes with very likely ranges for (**j**) Arctic sea ice extent change for September; (**k**) Arctic snow cover change for June (land areas north of 60° N); and (**l**) Change in near-surface (within 3–4 m) permafrost area in the Northern Hemisphere." Source: Reprinted from (IPCC 2019, pp. 17, 44); used with permission. Figure SPM.1 from IPCC 2019 Summary for Policymakers in IPCC Special Report on the Ocean and Cryosphere in a Changing Climate.

Higher acidity affects the balance of minerals in the water, which, for example, can make it more difficult for marine animals building their protective skeletons or shells. Some studies show the impact of ocean acidification on food chains and biodiversity, "but more efforts are required to strengthen our knowledge about the impact of acidification on the wider food web", states the IPCC (2019).

The ocean is continuing to acidify in response to ongoing ocean carbon uptake. The open ocean surface water pH is observed to be declining (virtually certain) by a very likely range of 0.017–0.027 pH units per decade since the late 1980s across individual time series observations longer than 15 years. The anthropogenic pH signal is very likely to have emerged for three-quarters of the near-surface open ocean prior to 1950 and it is very likely that over 95% of the near surface open ocean has already been affected. These changes in pH have reduced the stability of mineral forms of calcium carbonate due to a lowering of carbonate ion concentrations, most notably in the upwelling and high-latitude regions of the ocean. (IPCC 2019, p. 59)

- Ocean deoxygenation

Dissolved oceanic oxygen supports the largest ecosystems on the planet. Global warming impacts ocean oxygen in two ways: firstly, warmer water has a reduced capacity to hold oxygen and, secondly, the reduction in ocean mixing and circulation limits the uptake of oxygen from the atmosphere, because when water is not mixed, the top layer will be saturated with oxygen, while the bottom becomes anoxic. The oxygen in the ocean also depends on oxygen producing organisms living in the ocean. These organisms also depend on the water temperature and light availability and are influenced by climate change. Deoxygenation disrupts marine ecosystems causing loss of habitats and biodiversity, which can have knock-on effects, such as harming natural fish stocks and aquaculture.

There is a growing consensus that the open ocean is losing oxygen overall with a very likely loss of 0.5–3.3% over the period 1970 to 2010 from the ocean surface to 1000 m. Globally, the oxygen loss due to warming is reinforced by other processes associated with ocean physics and biogeochemistry, which cause the majority of the observed oxygen decline. The oxygen minimum zones (OMZs) are expanding by a very likely range of 3–8%, most notably in the tropical oceans, but there is substantial decadal variability that affects the attribution of the overall oxygen declines to human activity in tropical regions. Ocean model simulations predict a very likely decline in the dissolved oxygen content of the ocean by 3.2–3.7% (high emission scenario) by 2081–2100, relative to 2006–2015, or by 1.6–2.0% for the low emission scenario (IPCC 2019). The volume of the oceans OMZ is projected to grow by a very likely range of $7.0 \pm 5.6\%$ by 2100 during the high emission scenario, relative to 1850–1900 caused by a combination of a warming-induced decline in oxygen solubility and reduced ventilation of the deep ocean (IPCC 2019).

Deoxygenation accompanies ocean warming and ocean acidification as one of the three major oceanic consequences of rising atmospheric CO_2 levels (Levin and Breitburg 2015).

The decline in the oceanic oxygen content can affect ocean nutrient cycles and the marine habitat, with potentially detrimental consequences for fisheries, ecosystems and coastal economies. Oxygen loss is closely related to ocean warming and acidification caused by CO_2 increase driven by CO_2 emissions as well as biogeochemical consequences related to anthropogenic fertilization of the ocean. A combined effort investigating the different stressors will be most beneficial to understand future ocean changes (Schmidtko et al. 2017).

- Marine heatwaves

Marine heatwaves are periods of extremely high ocean temperatures. Scientists found that marine heatwaves have doubled in frequency and have become longer-lasting and more intense. The IPCC reports that marine heatwaves "have negatively impacted marine organisms and ecosystems in all ocean basins over the last two decades, including critical foundation species such as corals, sea grasses and kelps" (IPCC 2019). However, marine heatwaves are projected to further increase in frequency, duration, spatial extent and intensity (maximum temperature) (Frölicher et al. 2018). The IPCC reports that "climate models project increases in the frequency of marine heatwaves by 2081–2100, relative to 1850–1900, by approximately 50 times under a high emission scenario and 20 times under the low emission scenario" (IPCC 2019). The largest increases in frequency are projected for the Arctic and the tropical oceans. The intensity of marine heatwaves is projected to increase about 10-fold under a high emission scenario by 2081–2100, relative to 1850–1900 (IPCC 2019).

> In the absence of more ambitious adaptation efforts compared to today, and under current trends of increasing exposure and vulnerability of coastal communities, risks, such as erosion and land loss, flooding, salinization, and cascading impacts due to mean sea level rise and extreme events are projected to significantly increase throughout this century under all greenhouse gas emissions scenarios (very high confidence). Under the same assumptions, annual coastal flood damages are projected to increase by 2–3 orders of magnitude by 2100 compared to today (high confidence). (IPCC 2019)

- Extreme events

Projections in the IPCC show that climate change influences extreme events. Climate change is even projected to potentially cause abrupt changes in the ocean and the cryosphere (IPCC 2019).

In the ocean, a possible abrupt change is associated with an interruption of the Atlantic Meridional Overturning Circulation (AMOC). The AMOC is an important component of global ocean circulation. A slowdown of the AMOC could have consequences around the world: rainfall in the Sahel region could reduce, hampering crop production; the summer monsoon in Asia could weaken; increase in regional sea level around the Atlantic, especially along the northeast coast of North America; and there might be more winter storms in Europe (IPCC 2019).

- Impacts on marine ecosystems

The ocean is home to at least 230,000 known species in a variety of habitats that stretch from the flat coastline to the deep sea. Despite the ocean's important role in the climate system, its biodiversity serves, among other things, as an important food source. Changing physics and biochemistry of the ocean change the marine ecosystems and its services to humans. Marine biodiversity includes organisms that live in suspension in the water column with limited mobility (plankton), animals that live in the water column that can actively swim (nekton) and organisms that live within or on the sea floor (benthos). Moreover, most living phyla have marine representatives, with sizes that range from the smallest (archaea and viruses) to the largest living beings (the blue whale) (e.g., European Marine Board EMB). They play different roles throughout their life, inhabiting different environments and providing different functions. Particularly, wild fish capture is an important ecosystem service, and its abundance depends, among other factors, on the climatic related ocean conditions.

The IPCC states that ocean warming has contributed to observed changes in the biogeography of organisms ranging from phytoplankton to marine mammals, consequently changing community composition, and in some cases, altering interactions between organisms (IPCC 2019; summary provided in Figure 5). The IPCC projections show that along with ocean warming and changes in net primary productivity during the 21st century, global marine animal biomass and the maximum potential catches of fish stocks will be reduced, although with regional differences in the direction and magnitude of changes (IPCC 2019).

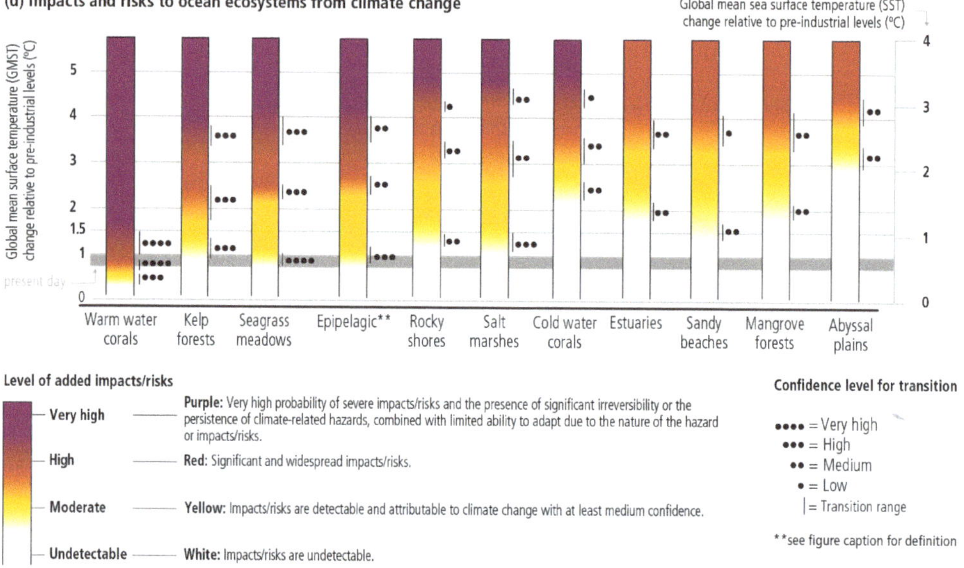

Figure 5. "Projected changes, impacts and for coastal and open ocean ecosystems. Specifically, assessment of risks for coastal and open ocean ecosystems based on observed and projected climate impacts on ecosystem structure, functioning and biodiversity. Impacts and risks are shown in relation to changes in Global Mean Surface Temperature (GMST) relative to pre-industrial level. Since assessments of risks and impacts are based on global mean Sea Surface Temperature (SST), the corresponding SST levels are shown." Source: Reprinted from (IPCC 2019, p. 33); used with permission. Figure SPM.3 (d) from IPCC 2019 Summary for Policymakers in IPCC Special Report on the Ocean and Cryosphere in a Changing Climate.

The projections from the IPCC show that these "future shifts in fish distribution and decreases in their abundance and fisheries catch potential due to climate change are projected to affect income, livelihoods, and food security of marine resource-dependent communities. Long-term loss and degradation of marine ecosystems compromise the ocean's role in cultural, recreational, and intrinsic values important for human identity and well-being" (IPCC 2019).

The projected redistribution of resources and abundance increases the risk of conflicts among fisheries, authorities or communities. Global warming compromises seafood safety through human exposure to elevated bioaccumulation of persistent organic pollutants and mercury in marine plants and animals, increasing prevalence of waterborne Vibrio pathogens, and heightened likelihood of harmful algal blooms. These risks are projected to be particularly large for human communities with high consumption of seafood, including coastal Indigenous communities, and for economic sectors such as fisheries, aquaculture, and tourism. (IPCC 2019, p. 26)

Higher acidity affects the balance of minerals in the water, which, for example, can make it more difficult for certain marine animals to build their protective skeletons or shells, as they must have access to available calcium in the seawater. This affects, for example, reef-building hard corals, as well as oysters, clams and snails that are composed of calcium carbonate (IPCC 2019). For coral polyps, for example, a 52–73% decline in larval settlement on reefs under lower pH levels has been shown (e.g., van Doorn et al. 2015). When the coral polyps have grown, scientists can also measure the calcification rates of hard corals. Studies show that ocean acidification has had a negative impact on the rate at which corals calcify, making them more brittle and less resilient to other factors influencing their survival in the future (e.g., van Doorn et al. 2015).

Coral reefs form habitats in the ocean, being home to the richest and most diverse biodiversity of our ocean. The survival of thousands of other marine species depends on coral reefs—many of which we rely on for food. In addition, they act as buffers against sea-level rise and increased storm intensity and thus play a critical role in mitigating and adapting to climate change (IPCC 2019).

- Impacts on coastal communities

Coastal areas are particularly vulnerable to the effects of climate change, such as sea level rise, storm intensity and flooding, increased temperatures, ocean acidification and low oxygen zones, all resulting in changing marine ecosystems. At the same time, coastal areas are highly populated, and today, about two-thirds of the world's population live within 60 km of the coast (United Nations Atlas of Oceans n.d.), and societies depend on global transport via shipping routes and harbors (e.g., IPCC 2019). More than 600 million people (around 10 per cent of the world's population) live in coastal areas that are less than 10 m above sea level (IPCC 2019). The IPCC summarizes that "coastal areas are zones of concentrated biodiversity and natural productivity and will be particularly affected by multiple stressors because this is where most human activities take place and where pressure accumulates" (IPCC 2019).

The IPCC report states that increased mean and extreme sea level, ocean warming and ocean acidification will be the major threats for communities in low-lying coastal areas, delta regions and resource rich coastal until 2050 under current adaptation

(IPCC 2019). At global scale, coastal protection could reduce flood risk by 2–3 orders of magnitude during the 21st century, but depends on large investments, and some island communities have already lost their homes due to sea level rise (IPCC 2019). Such investments can be cost efficient for densely populated urban areas, but they might be difficult to afford for rural and poorer areas (IPCC 2019).

5. Conclusions

Climate change is among the main drivers of change also for the ocean system. At the same time, the ocean is part of the climate system and provides "memory" and regional redistribution of climate signals. In the 2030 Agenda for sustainable development, both climate and ocean have specific goals (SDG 13 and 14, respectively). While both the ocean and climate issues transgress many of the goals (e.g., Nilsson et al. 2016), the rapidly advancing scientific understanding of both the ocean and climate system has allowed for the in-depth assessment of what is known (e.g., IPCC 2019, extensively referred to in this paper).

On the other hand, the 2030 Agenda for Sustainable Development and the Paris Climate Agreement are focused on action. While the scientific evidence and problem diagnostic down to the regional level is well advanced, the impacts on the local level and the assessment of development options and solutions to adapt and mitigate climate and ocean change are less well researched. A recent report from the High Level Panel for Sustainable Ocean Economy discusses "The Ocean as a Solution for Climate Change: 5 Opportunities for Action" (Hoegh-Guldberg et al. 2019) and suggests that up to 25% of the needed CO_2 emission reductions can be obtained by ocean-related actions, which include innovation and efficiency gains in the maritime industry but also nature-based solutions to increase ocean uptake of CO_2 and ocean-based energy production.

Connecting Ocean Science with the quest for action and solutions is the focus of the upcoming UN Decade of Ocean Sciences for Sustainable Development (e.g., Ryabinin et al. 2019). In order to fully achieve the Ocean Decade Objectives as well as the ocean dimension of the 2030 Agenda, growth and transformation in how we conduct ocean science towards more integration across disciplines and knowledge systems are essential (Pendleton et al. 2020). This will require a sustained ocean observation system; all ocean data shared freely; new and more effective ways of analyzing observational data fused with ocean and climate model; and enhancing timely assessment, predictions and scenario development of future ocean conditions (Visbeck 2018). At the same time, we need to grow ocean science capacity and capabilities worldwide and establish the sharing of resources and information.

A particular focus is the countries of the global south and small island states who are fully aware of the challenges. Ocean science must come together to be in a position to support decision makers by providing knowledge and frameworks to weigh the ecological, environmental and human impacts of different sustainable development pathways. New and innovative ways of collaborating amongst all ocean stakeholders will need to be identified. Disciplines and perspectives not always represented in ocean science will need to be involved in order to share their knowledge and attain a more holistic understanding.

Author Contributions: S.K. and M.V. contributed equally to the review article.

Conflicts of Interest: The authors declare no conflict of interest.

References

European Marine Board (EMB). 2019. *Navigating the Future V: Marine Science for a Sustainable Future. Position Paper 24 of the European Marine Board.* Ostend: European Marine Board, ISBN 9789492043757. [CrossRef]

Frölicher, Thomas L., Erich M. Fischer, and Nicolas Gruber. 2018. Marine heatwaves under global warming. *Nature.* [CrossRef]

Hoegh-Guldberg, Ove, Ken Caldeira, Thierry Chopin, Steve Gaines, Peter Haugan, Mark Hemer, Jennifer Howard, Manaswita Konar, Dorte Krause-Jensen, Elizabeth Lindstad, and et al. 2019. *The Ocean as a Solution to Climate Change: Five Opportunities for Action.* Report. Washington, DC: World Resources Institute, Available online: http://www.oceanpanel.org/climate (accessed on 8 February 2021).

IPCC. 2014. *Climate Change 2014: Synthesis Report. Contribution of Working Groups I, II and III to the Fifth Assessment Report of the Intergovernmental Panel on Climate Change.* Edited by Rajendra K. Pachauri and Leo Meyer. Geneva: IPCC, 151p.

IPCC. 2018. Summary for Policymakers. In *Global Warming of 1.5 °C. An IPCC Special Report on the Impacts of Global Warming of 1.5 °C Above Pre-Industrial Levels and Related Global Greenhouse Gas Emission Pathways, in the Context of Strengthening the Global Response to the Threat of Climate Change, Sustainable Development, and Efforts to Eradicate Poverty.* Edited by Valerie Masson-Delmotte, Panmao Zhai, Hans-Otto Pörtner, Debra C. Roberts, James Skea, Priyadarshi R. Shukla, Anna Pirani, Wilfran Moufouma-Okia, Clotilde Péan, Roz Pidcock and et al. Geneva: World Meteorological Organization, p. 32.

IPCC. 2019. *IPCC Special Report on the Ocean and Cryosphere in a Changing Climate.* Edited by Hans-Otto Pörtner, Debra C. Roberts, Valerie Masson-Delmotte, Panmao Zhai, Melinda Tignor, Elvira Poloczanska, Katja Mintenbeck, André Alegría, Maike Nicolai, Andrew Okem and et al. Geneva: World Meteorological Organization, in press.

Levin, Lisa A., and Denise L. Breitburg. 2015. Linking coasts and seas to address ocean deoxygenation. *Nature Climate Change* 5: 401–3. [CrossRef]

. Meredith, Michael P. 2019. The global importance of the Southern Ocean, and the key role of its freshwater cycle. *Ocean Challenge* 23: 27–32.

Nilsson, Måns, Dave Griggs, and Martin Visbeck. 2016. Policy: Map the interactions between Sustainable Development Goals. *Nature* 534: 320–2. [CrossRef] [PubMed]

Pendleton, Linwood, Karen Evans, and Martin Visbeck. 2020. We need a global movement to transform ocean science for a better world. *Proceedings of the National Academy of Sciences of the United States* 117: 9652–55. [CrossRef] [PubMed]

Rockström, Johan, Will Steffen, Kevin Noone, Åsa Persson, F. Stuart Chapin III, Eric Lambin, Timothy M. Lenton, Marten Scheffer, Carl Folke, Hans Joachim Schellnhuber, and et al. 2009. Planetary Boundaries: Exploring the safe operating space for humanity. *Ecology and Society* 14: 32. Available online: http://www.ecologyandsociety.org/vol14/iss2/art32/ (accessed on 8 February 2021).

Ryabinin, Vladimir, Julian Barbière, Peter Haugan, Gunnar Kullenberg, Neville Smith, Craig McLean, Ariel Troisi, Albert Fischer, Salvatore Aricò, Thorkild Aarup, Peter Pissierssens, and et al. 2019. The UN Decade of Ocean Science for Sustainable Development. *Frontiers in Marine Science* 6. [CrossRef]

Schmidt, Stefanie, Barbara Neumann, Ywonne Waweru, Carole Durussel, Sebastian Unger, and Martin Visbeck. 2017. SDG 14—Conserve and Sustainable Use the Oceans, Seas and Marine Resources for Sustainable Development. In *A Guide to SDG Interactions: From Science to Implementation*. Edited by Griggs David, Nilsson Måns, Stevance Anne-Sophie and McCollum David. Paris: International Council for Science (ICSU), pp. 174–218.

Schmidtko, Sunke, Lothar Stramma, and Martin Visbeck. 2017. Decline in global oceanic oxygen content during the past five decades. *Nature* 542: 335–9. [CrossRef] [PubMed]

Schmitt, Raymond W. 2018. The ocean's role in climate. *Oceanography* 31: 32–40. [CrossRef]

Talley, Lynne D. 2013. Closure of the global overturning circulation through the Indian, Pacific, and Southern Oceans: Schematics and transports. *Oceanography* 26: 80–97. [CrossRef]

United Nations Atlas of Oceans. n.d. Available online: www.oceansatlas.org (accessed on 28 June 2020).

van Doorn, Erik, Rene Friedland, Uwe Jenisch, Ulrike Kronfeld-Goharani, Stephan Lutter, Konrad Ott, Martin Quaas, Katrin Rehdanz, Wilfried Rickels, Jörn Schmidt, and et al. 2015. *World Ocean Review 2015: Living with the Oceans 4. Sustainable Use of our Oceans—Making Ideas Work*. Hamburg: Maribus gGmbH, ISBN 978-3-86648-253-1.

Visbeck, Martin. 2018. Ocean science research is key for a sustainable future. *Nature Communications* 9: 690. [CrossRef]

Visbeck, Martin, Ulrike Kronfeld-Goharani, Barbara Neumann, Wilfried Rickels, Jörn Schmidt, Erik van Doorn, Nele Matz-Lück, Konrad Ott, and Martin F. Quaas. 2014. Quaas Securing blue wealth: The need for a special sustainable development goal for the ocean and coasts. *Marine Policy* 48: 184–91. [CrossRef]

Deep-Sea Mining: Can It Contribute to Sustainable Development?

Luise Heinrich and Andrea Koschinsky

1. Introduction

Sustainable development is a kind of "development that meets the needs of the present without compromising on the ability of future generations to meet their own needs" (Brundtland 1987, para. 27). It aims at balancing economic development with human well-being and environmental conservation, taking into account concerns of inter- and intragenerational equity. The need to divert from a business-as-usual development path to a more sustainable one was re-emphasized by the international community in 2015, when all of the United Nations' member states adopted the 2030 Agenda for Sustainable Development (A/RES/70/1). The 2030 Agenda presents "a plan of action for people, planet and prosperity" (A/RES/70/1, preamble) and brings together the 2000 Millennium Development Goals (MDGs) and the climate and environment agenda rooted in the 1992 Earth Summit (Rio de Janeiro, Brazil) (BMU 2015). At the center of the 2030 Agenda are 17 interlinked sustainable development goals (SDGs) with 169 associated targets, which reflect the 2030 Agenda's objectives to "end poverty and hunger everywhere; to combat inequalities; to protect human rights and promote gender equality and the empowerment of women and girls; and to ensure the lasting protection of the planet and its natural resources", as well as the creation of "conditions for sustainable, inclusive and sustained economic growth, shared prosperity and decent work for all, taking into account different levels of national development and capacities" (A/RES/70/1, page 3).

Whether mining is compatible with the concept of sustainable development is debatable. On the one hand, mineral resources serve as important raw materials used for the manufacture of a myriad of goods, including, inter alia, construction materials and electronic devices (UNDP and UN Environment 2018). Furthermore, the export of mineral raw materials makes up a large share of the national economies of many countries. On the other hand, mining entails the exploitation of a finite resource which is often associated with substantial environmental destruction. Furthermore, once depleted, the resource will no longer be available for future generations, as mineral deposits take millions of years to form. Due to declining ore grades, it is likely that terrestrial mines will in the future be forced to expand more rapidly both laterally and vertically to keep the production constant. Furthermore, it is expected that mines will move into more remote terrains, which taken altogether will likely intensify social and environmental pressures (Calvo et al. 2016).

Deep-sea mining, which describes the recovery of marine minerals from the deep seabed, may in the future contribute to meeting the metal demand of the growing world population (Hein et al. 2013). The idea of deep-sea mining first emerged in the 1960s, when the economic potential of marine mineral resources was widely recognized (Mero 1965; Sparenberg 2019). At that time, the interest in deep-sea mining was purely economic and geostrategic, as deep-sea mining was seen as a means to generate revenue and to decrease the dependency on foreign metal exports (Sparenberg 2019; Koschinsky et al. 2018). For a long time, the deep-sea mining narrative has, in this regard, followed the assumption that marine mineral resources are of greater value if they are exploited and converted into revenue (Christiansen et al. 2019). This is underpinned by the claim that deep-sea mining could provide the metals needed for the transition to a low-carbon economy (Paulinkas et al. 2020; Hein et al. 2013). Moreover, studies claim that deep-sea mining may, in fact, be more environmentally friendly than terrestrial mining (Paulinkas et al. 2020; Batker and Schmidt 2015; Hein and Koschinsky 2014; Koschinsky et al. 2018). This rather positive outlook on deep-sea mining is, however, increasingly challenged, as concerns about the potential large-scale and long-term environmental impacts and the potential implications for human and ecosystem well-being are raised (Weaver and Billet 2019). Furthermore, it has been questioned whether a comparison of terrestrial and deep-sea mining is even warranted, given that there is no indication that deep-sea mining will eventually replace terrestrial mining. Instead, it is more likely that both will be carried out in parallel, ultimately intensifying environmental and social conflicts even further (Christiansen et al. 2019).

With commercial deep-sea mining on the horizon, it becomes increasingly important to explore if and how deep-sea mining can contribute to sustainable development. This requires a thorough assessment of environmental, economic and social concerns (Figure 1). Following this introduction, this chapter will present the three different types of marine mineral deposits under consideration to be mined, including envisioned mining concepts, and quickly explain the legal context of deep-sea mining. Subsequently, the chapter will outline environmental, economic and social considerations and conclude with a section on implications for sustainable development.

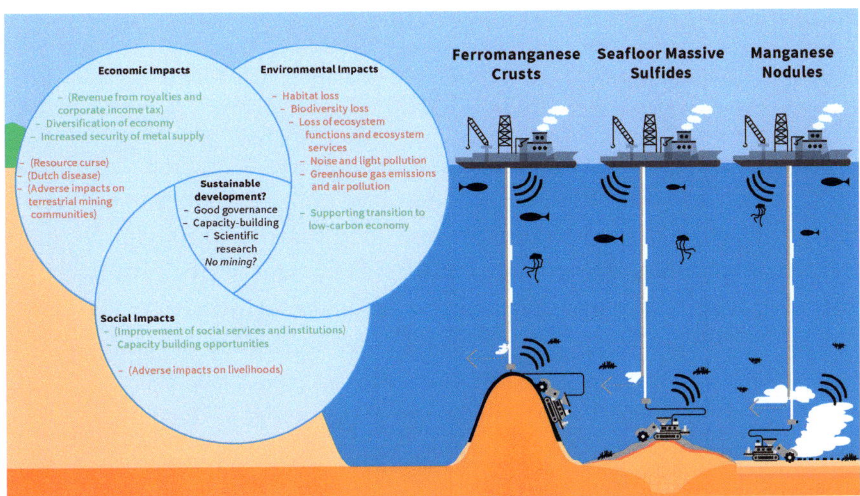

Figure 1. Overview of marine mineral deposits, mining techniques, and impacts. Positive and negative impacts are shown in green and red, respectively. Impacts in parentheses indicate potential impacts, which can be good or bad depending on external factors, such as the availability of effective policies or capacity-building initiatives. Impacts without parentheses are certain. Source: Figure by author; modeled after (Aldred 2019).

2. Types of Marine Mineral Deposits

Manganese nodules (here forth simply referred to as nodules) are small, potato-shaped mineral concretions, which mainly consist of concentric intergrown layers of iron and manganese oxides, but also contain significant quantities of various metals, including nickel, copper, cobalt, molybdenum, zinc, platinum, tellurium, and rare earth elements (Hein and Koschinsky 2014). They form by the precipitation of metals from seawater or sediment pore water and occur nearly everywhere on the world's oceans, but are especially abundant in the Clarion-Clipperton-Zone (CCZ), the Peru Basin, near the Cook Islands (all located in the Pacific Ocean), and the Central Indian Ocean Basin (Hein et al. 2013; Petersen et al. 2016). Most nodule mining concepts envision mining operations to consist of one or more remotely operated vehicles, which will collect nodules at the seafloor. From there, the nodules will be pumped through a riser pipe and deposited onboard a production support vessel at the surface. Onboard, the nodules will be washed, partially dried and stored until they are collected by a transport vessel and brought to land, where they will be metallurgically processed. The wastewater sediment mixture will be returned to the water column (Atmanand and Ramadass 2017; Blue Mining 2014; Hong et al. 2010; Ramboll IMS & HWWI 2016). It has been suggested that this should happen at near-seafloor depth to avoid the contamination of pelagic ecosystems (Drazen et al. 2020).

Ferromanganese crusts (here forth simply referred to as crusts) form through the precipitation of metals on the sediment-free summits, platforms, slopes and saddles of seamounts in water depth between 400 and 7000 m over the course of millions of years (Hein and Koschinsky 2014). They consist of strongly intergrown sub-crystalline iron and manganese oxide layers of up to 25 cm thickness and contain economically interesting quantities of other metals, including nickel, copper, cobalt, molybdenum, zirconium, niobium and rare earth elements and reach a known maximum thickness of about 25 cm (Halbach et al. 1982; Hein et al. 1992; Lusty et al. 2018). It is believed that there are thousands of seamounts located across the world's oceans, but the Prime Crust Zone (PCZ), which stretches from the Mariana Trench to the Hawaiian Islands, is of particular interest because of its high abundance of crusts with highly valuable metal contents (Wessel et al. 2010; Lusty et al. 2018; Hein and Koschinsky 2014). Due to their firm attachment to the underlying rock, the mining of crusts is considered challenging (Lusty et al. 2018; Koschinsky et al. 2018). In August 2020, the Japan Oil, Gas, and Mineral National Corporation (JOGMEC) announced that it conducted the world's first successful crust-mining test, during which they excavated 649 kg of crusts from the seafloor off the Japanese coast, using a crust-excavating testing machine developed by JOGMEC itself (JOGMEC 2020).

Seafloor massive sulfide (SMS) deposits form in hydrothermally active areas through the precipitation of minerals, when hot metal-rich hydrothermal fluids cool or get in contact with cold ambient seawater (Hannington et al. 2005). They consist mainly of metal–sulfur compounds and contain significant amounts of iron, copper, zinc, silver, and gold, as well as smaller quantities of rare earth elements (Monecke et al. 2014). SMS deposits are located in geologically active areas such as mid-ocean ridges, and in volcanic arc and back arc basins, and at intraplate volcanoes (Petersen et al. 2016). Based on plume studies and deposit occurrence models, Hannington et al. (2011) estimated that there are between 500 and 5000 vent fields with associated mineral deposits. Hydrothermal vent fields are considered active, while the venting of hydrothermal fluids is ongoing, inactive and eventually extinct when it ceases. Vents located on slow-spreading ridges (e.g., Atlantic Ocean) can last for hundreds of thousands of years whereas those located on fast-spreading riches (e.g., East Pacific Rise) often rise and fall over decades (Copley et al. 2016).

Deep-sea mining of seafloor massive sulfide deposits will likely concentrate on inactive vent sites, which have accumulated over a longer time than active vent sites (German et al. 2016; Van Dover et al. 2018). Furthermore, active venting of hot hydrothermal fluids may pose a significant threat to mining equipment (SPC 2013c). Mining concepts currently envision the combined use of different seafloor vehicles (bulk cutter, auxiliary cutter and collector), which will cut and collect the ore at the seafloor. From there, it will be pumped to the seafloor, cleaned from sediment onboard a mining vessel and then transported to shore for further metallurgical processing (SPC 2013c). More recently, the use of vertical cutter systems has been suggested (Spagnoli et al. 2016).

3. Deep-Sea Mining in Areas within and Beyond the Limits of National Jurisdiction

The responsibility of regulating the exploration and exploitation of marine mineral deposits in territorial waters, exclusive economic zones (EEZs) and the continental shelf zones lies with the respective coastal states, who are obligated by the United Nations Convention on the Law of the Sea (UNCLOS) to adopt appropriate regulations that are "no less effective than international rules, standards and recommended practices and procedures" (UNCLOS, Article 208 (3), see Section 4.3 below for information on environmental obligations of coastal states). Deep-sea mining in areas beyond national jurisdiction is primarily regulated by Part XI of UNCLOS (the Area) and the corresponding 1994 Agreement relating to the implementation of Part XI of UNCLOS (1994 IA). The international seabed (termed the Area by UNCLOS) and its resources constitute the Common Heritage of Mankind (CHM) (UNCLOS, Article 136), which means that the resources of the Area are vested in mankind as a whole (UNCLOS, Article 137 (1)), effectively prohibiting states from claiming, acquiring, or exercising sovereign rights over them (UNCLOS, Article 137 (3)). Instead, the resources of the Area are managed by the International Seabed Authority (ISA), which has been established by UNCLOS (153 (1)), and is to act on behalf of mankind as a whole (UNCLOS, 137 (2)).

The CHM principle has been established to ensure that the benefits from exploiting the resources of the Area are shared by all countries "irrespective of the geographic location of States, whether coastal or land-locked, and taking into particular consideration the interests and needs of developing States" (UNCLOS, Article 140). As such, its objective is to prevent a situation in which the benefits obtained from seabed mining can only be enjoyed by industrialized countries, which have the financial capacity and technical skill to carry out such an expensive and risky endeavor (Jaeckel et al. 2016). Key elements of the CHM principle include (1) the exclusive use of the international seabed for peaceful purposes (UNCLOS, Article 141), (2) the principle of non-appropriation (UNCLOS Article 137 (1)), (3) the reservation of mineable areas for developing states in the Area, (4) the equitable sharing of any monetary and non-monetary benefits (UNCLOS, Article 140(2)) and (5) the protection and preservation of the marine environment for the benefit of current and future generations (UNCLOS, Article 145). To this end, the ISA's main tasks include the development of a regulatory and administrative structure that allows the sharing of monetary and non-monetary benefits and the development of stringent environmental regulation, which ensures the protection and preservation of the marine environment from the impacts of deep-sea mining, taking into account concerns of intergenerational and intragenerational equity (Frakes 2003; Jaeckel et al. 2016; Bourrel et al. 2018; Joyner 1986; Kiss 1985).

Deep-sea mining in the Area can either be carried out by the Enterprise (the ISA's would-be mining entity responsible for mining, transporting, processing and marketing marine minerals recovered from the Area) and, in association with the ISA, by member states of UNCLOS, state and private enterprises, natural or juridical persons who have the nationality of a member state and who are sponsored by such a state (UNCLOS Article 139). The sponsoring state is required to ensure that the contractor (i.e., the entity entering into exploration or mining contracts with the ISA) complies with the terms of its contract and with the relevant provisions of international law. In this regard, the sponsoring state has an obligation of due diligence in setting and enforcing its laws and regulations, meaning that it has to adopt, implement and enforce appropriate rules and regulations (ITLOS 2011), which, according to Lily (2018), may include the provision of "institutional capabilities such as an identified regulatory body, with monitoring and enforcement functions and access to appropriate personnel, equipment and other technical capacity to implement them"(p. 2). Wherever sponsoring states have implemented appropriate measures, they cannot be held liable for a contractor's misconduct (ITLOS 2011).

As of December 2020, the ISA has entered into 30 exploration contracts, eighteen of which are for nodules, five for crusts and seven for SMS deposits (ISA 2020).

4. Environmental Considerations

4.1. Environmental Impacts of Deep-Sea Mining

4.1.1. Biological Impacts

Manganese nodules are loosely placed in and on top of the sediment of the abyssal plains of the oceans in an environment, which is characterized by high pressure, low temperature and very slow dynamics of (bio)geochemical processes. The nodules serve as a habitat for a variety of sessile and mobile faunal taxa (e.g., bacteria, nematodes, harpacticoid copepods, polychaeta, isopod crustaceans, holothurians, fish, corals, bryozoans, xenophyophores, and sponges), which typically feed on detritus and fecal pellets produced by zooplankton sinking down from the sea surface (marine snow) (SPC 2013b; Vanreusel et al. 2016; Weaver and Billet 2019; Amon et al. 2016). Collector vehicles moving over the seafloor will not only destroy the nodules and with it the habitat for organisms using the nodules as hard substrate, but will also stir up the sediment, effectively threatening bottom-dwelling and filter-feeding organisms (Weaver and Billet 2019; Koschinsky et al. 2018). In addition to this, the re-deposition of the suspended sediment is also expected to adversely affect these organisms, as this would likely happen at a much higher rate than natural sedimentation (Weaver and Billet 2019).

Ferromanganese crusts provide solid substrate for sessile filter feeding taxa (e.g., corals, sponges) and a variety of mobile taxa, including echinoderms, squids, and foraminifera (Mullingneaux 1987; Weaver and Billet 2019; Clark et al. 2010). The distribution of species and the composition of communities vary depending on factors like water depth, current flow and type of substrate (Clark et al. 2010). Research has indicated that the seamounts host considerably more biomass than the slopes of continental margins at the same depth (Rowden et al. 2010). The removal of the crusts would inevitably lead to the vast destruction of large areas of habitat. Furthermore, the mining of crusts could produce particle plumes, including resuspended sediment and abraded crust particles. However, as seamounts will only accumulate sediment on plateaus and in crevices, the size and distribution of the particle plume will likely be much smaller than the plume generated by nodule mining (SPC 2013b; Koschinsky et al. 2018; Hein and Koschinsky 2014).

SMS deposits, specifically active hydrothermal vent fields, provide unique habitats for a variety of highly specialized organisms (e.g., shrimp, tube worms and bacteria) (SPC 2013c). Many of these species are endemic to individual vents and rely on a well-functioning symbiotic relationship with certain chemoautotrophic species (SPC 2013b; Van Dover et al. 2018). Vent communities also show a zonation, meaning that the different organisms occur at different distances to the vent (Rogers et al. 2012). The impacts of SMS mining will likely be site-specific due to variations in local abiotic conditions, including substrate type, water depth, temperature, salinity and particulate organic matter supply from the surface (Boschen et al. 2016). Overall, the area affected by mining will be smaller than the area influenced by nodule or crust mining, as SMS mines would mostly extent into the sub-seafloor (SPC 2013c; Weaver and Billet 2019). However, due to the uniqueness of individual active vent habitats, the mining of active vents would risk destroying rare types of habitat. Furthermore, due to the smaller size of the deposits, more vent sites would likely have to be mined. However, it is more likely that inactive vent sites would be preferentially mined in the future, as they may provide larger ore deposits and would be technically easier to mine than active vent sites. While here fauna can be expected to be more similar to the ambient deep-sea fauna of the region, as the typical vent fauna can only survive at actively venting sites, the paucity of ecological studies at inactive SMS deposits makes clear assessments of a potential environmental impact of mining difficult (Van Dover 2019). Like for active SMS, the affected area of mining would be much smaller than the affected area of nodule or crust mining.

4.1.2. Geochemical Impacts

Deep-sea mining can also cause geochemical changes by altering the chemical equilibrium of the sediment-water interface as a consequence of the excavation of marine mineral resources and the removal of surface sediment. In the case of nodule mining, the extent of the release of toxic metals from seawater and sediment pore water is believed to be small, unless mining causes particularly deep disturbances. Strong interferences could, however, occur in areas where the oxygen penetration depth in the sediment is very low. Recent studies suggest, however, that oxygen reaches depths of more than 1.5 m throughout the CCZ (Mewes et al. 2014; Volz et al. 2020). In the Peru Basin, where nodules are also highly abundant, the oxygen penetration depth is only between 10–15 cm (Haeckel et al. 2001; Paul et al. 2018). Crust mining is not expected to cause a significant release of toxic metals, as the crusts typically form under fully oxic conditions. However, if crusts on shallow seamounts close to the oxygen minimum zone would be mined, a partial redissolution of manganese oxide from crust particles and release of trace metals within the oxygen minimum zone could take place (Koschinsky et al. 2003). The mining of SMS deposits may have a substantial geochemical impact because of the high oxidation potential and reduced state of the sulfide minerals (Van Dover et al. 2020). Research has shown that even species inhabiting active vent sites, which are characterized by a comparatively high concentration of metals in the surrounding water, may be negatively affected by elevated metal concentrations due to mining (e.g., Hauton et al. 2017). Although many vent species may be more adapted to changing environmental conditions and appear to have developed mitigation strategies against metal toxicity (vent mussels, for example, store immobile metal compounds in their tissue, Koschinsky 2016), it is unclear to which limits these adaptation strategies would protect these organisms against metal release from SMS mining.

4.1.3. Particle Plumes

The operation of the collector vehicles at the seafloor and the discharge of excess sediment and water from the mining vessel will create metal-rich particle plumes close to the seafloor and in the water column, which may negatively affect benthic and pelagic ecosystems and may extend far beyond the mine site (SPC 2013a, 2013b, 2013c). Whereas early research mostly relied on hydrodynamic models to anticipate the dispersion of the plume (Jankowski and Zielke 2001; Rolinski et al. 2001), more recent experiments show aggregation effects, indicating that previous research may have overestimated the range of dispersion of the plume (Gillard et al. 2019). Nevertheless, fine particles can be transported over long distances and potentially negatively affect marine organisms (Weaver et al. 2018). The mining of the slopes of seamounts and active vent sites is not expected to produce large particle plumes, as these are generally not covered with a thick sediment layer. Guyots and crevices of seamounts, as well as inactive vent sites can, however, accumulate sediment. Similarly, inactive hydrothermal vent sites may also be covered by several centimeters of sediment, which may be dispersed during mining and the discharge of excess water and sediment from the mining vessel (Weaver and Billet 2019; Van Dover et al. 2020).

4.1.4. Noise and Light Pollution

Exposure to noise and vibrations resulting from mining operations can compromise the ability of marine organisms to communicate and to detect prey. As noise travels well underwater, noise pollution could affect an area much greater than the mine site (Weaver et al. 2018). Noise impacts may be particularly severe in water depth in the upper 2000 m of the water column, where it may negatively affect marine mammals (Weaver and Billet 2019). Similarly, lights attached to mining equipment could disturb species that are accustomed to living in a dark environment (Popper et al. 2003; Weaver et al. 2018; Weaver and Billet 2019). Furthermore, artificial light may conceal bioluminescence, which may compromise the ability of marine organisms to navigate, mate, detect food and defend against predation. Near the vessels, artificial light may also attract organisms and disrupt their movement and above the sea surface. Furthermore, birds may be adversely affected by the lights illuminating the working decks of the mining vessels (Weaver and Billet 2019).

4.1.5. Greenhouse Gas Emissions and Air Pollution

The combustion of fuel oil onboard the mining and transport vessels will cause the release of greenhouse gases (i.e., CO_2, CH_4, N_2O) and other air pollutants (e.g., CO, SO_x, NO_x, NMVOCs, PM) (IMO 2015). These emissions will contribute to global warming, acidification, and the formation of photochemical ozone (Huijbregts et al. 2016). Thus far, the impacts to air directly resulting from deep-sea mining have received little attention in research. They should, however, be considered in a holistic assessment of the environmental impacts caused by deep-sea mining, especially in the context of climate change mitigation, and incorporated in regulatory frameworks (Heinrich et al. 2020).

4.1.6. Ecosystem Services

The impacts caused by deep-sea mining may also affect ecosystem functions and services (Le et al. 2017; Orcutt et al. 2020; Thornborough et al. 2019). Ecosystem functions of marine ecosystems include element and nutrient cycling, the provision of breeding grounds, nursery habitats and refugia, bioturbation, dispersal and connectivity, as well as primary and secondary productivity, metabolic activity and respiration (Le et al. 2017). Ecosystem services describe the benefits humans obtain from well-functioning ecosystems and are commonly subdivided into provisioning services, regulating services, supporting services and cultural services (MEA 2005). Provisioning services obtained from marine ecosystems, for example, include fish, shellfish, biomaterials, pharmaceuticals and industrial agents. Regulating services, for example, include carbon sequestration, the control of pests and populations, and the storage, burial, transformation and detoxification of waste material and pollutants. Cultural services include aesthetic and spiritual value, educational services and the notion of ocean stewardship. Supporting services include the ecosystem functions listed above (Le et al. 2017; Armstrong et al. 2012). Biodiversity is considered to be of particular importance in supporting ecosystem functions, although the relationship between biodiversity and ecosystem services has not yet been fully understood (Balvanera et al. 2014; Bennett et al. 2015). How and to what extent deep-sea mining will affect ecosystem functions and ecosystem services is uncertain but may be substantial. It should, therefore, be considered in the development of regulatory frameworks and management practices (Thornborough et al. 2019; Le et al. 2017).

4.2. The Mitigation Hierarchy

The mitigation hierarchy provides a systematic approach for reacting to the environmental impacts of an activity. Its main objective is to avoid net loss of biodiversity and, wherever possible, to achieve net gain. The mitigation hierarchy requires the consideration of four elements in a strict hierarchical order: (1) avoid, (2) minimize, (3) restore, (4) compensate/offset (Billet et al. 2019). Although originally developed for application in a terrestrial setting, it is now increasingly applied to coastal and marine environments, including the deep-sea. The first objective of the mitigation hierarchy is to avoid deep-sea mining altogether by reducing the overall demand for metals through recycling, substituting non-renewable with renewable materials and changing consumer behavior, although it is unclear whether this would be sufficient to meet the increasing demand of the growing world population (Billet et al. 2019; Rühlemann et al. 2019). If the complete avoidance of deep-sea mining is, indeed, impossible, then measures should be undertaken to at least protect certain areas from mining through the establishment of marine protected areas in which no mining can take place. An important measure in this regard is the establishment of regional-scale environmental management plans (REMPs), which is supposed to help maintain regional biodiversity, ecosystem structures and ecosystem function and to preserve typical regional ecosystems (Cuvelier et al. 2018; Niner et al. 2018; Jacob et al. 2016). According to Jones et al. (2019), REMPs for deep-sea mining may include "an assessment of the probability, duration, frequency and reversibility of environmental impacts, the cumulative and transboundary impacts, the magnitude and spatial extent of the effects, the value and vulnerability of the area likely to be affected including those with protection status and the extent of uncertainty in any of the above" (p. 175). The ISA has, until now, only adopted a REMP for the nodule fields of the CCZ, whose central component is a network of nine Areas of Particular Environmental Interests (ISBA/24/C/3). The APEIs cover an area of 400 km × 400 km, representing the nine sub-regions of the CCZ. The guiding principles of the CCZ REMP are listed as (1) the CHM, (2) the precautionary approach, (3) the protection and preservation of the marine environment, (4) the requirement to conduct environmental impact assessments, (5) the conservation and sustainable use of biodiversity and (6) transparency. The establishment of representative APEIs is complicated by the persisting lack of knowledge about species abundances and community composition in the deep sea. There is, however, a clear call for the establishment of further REMPS (including APEIs) in the Area, including prospective sites for the mining of crusts and SMS deposits. The selection of APEIs should be guided and by a comprehensive set of environmental criteria and objectives.

Moreover, Tunnicliffe et al. (2020) point out that "clearly identified targets using well-defined and standardized performance indicators [are needed] to evaluate progress (or lack thereof) towards achieving desired outcomes" (p. 3). Due to the uniqueness of SMS habitats, finding representative sites for the placement of APEIs will, however, be challenging (Koschinsky et al. 2018). Within areas of national jurisdiction, the Pacific Community established the Regional Environmental Management Framework for Deep Sea Minerals Exploration and Exploitation in cooperation with the EU (Swaddling 2016).

The second objective of the mitigation hierarchy is to minimize adverse environmental impacts as much as possible via technological means. While habitat destruction by seafloor vehicles is inevitable in a deep-sea mining context, it may be possible to reduce the impact of the particle plume. Niner et al. (2018), for example, suggest the use of shrouds on seafloor vehicles to limit the production and spreading of fine particles and Cuvelier et al. (2018) mention the possibility to increase flocculation to encourage a faster settling of the plume. Furthermore, the use of alternative energy sources (e.g., liquefied natural gas (LNG)) and the increase of the energy efficiency of the ship engines could limit the release of greenhouse gases and air pollutants (Heinrich et al. 2020). The third objective of the mitigation hierarchy is to restore ecosystem function and services after destruction. While this is common practice in terrestrial mining, the restoration of deep-sea ecosystems is extremely difficult due to the large scale of the affected areas, persisting knowledge gaps, and limited economic feasibility (Van Dover et al. 2014; Niner et al. 2018; Billet et al. 2019). The compensation/offsetting of biodiversity loss can be considered as a last option to prevent a net loss of biodiversity. This can be achieved by protecting or restoring similar habitats to those mined (like for like), or to create new biodiversity of a different kind in different types of environments (out of kind). It may, furthermore, be possible to compensate in an entirely different manner, for example, through investing in capacity-building initiatives. However, Niner et al. (2018) point out that out of kind compensation can neither negate biodiversity loss nor compensate for lost ecosystem functions and should, therefore, not be considered true offsets.

4.3. Environmental Regulation

4.3.1. National Jurisdiction

In areas within national jurisdiction, UNCLOS obligates coastal states to ensure the protection and preservation of the marine environment (UNCLOS, Articles 192 and 193). In this regard, UNCLOS requires states to attempt "as far as practicable, directly or through the competent international organization to observe, measure, evaluate and analyze by recognized scientific methods, the risks or effects of pollution of the marine environment" resulting from activities "which they permit or in which they engage" (UNCLOS, Article 204). Wherever states suspect "substantial pollution [or] significant harmful changes to the marine environment", they are required to "as far as practicable, assess the potential effects of such activities on the marine environment and shall communicate reports of the results of such assessments" (UNCLOS, Article 206) to the competent international organizations (UNCOS, Article 205). With respect to deep-sea mining, coastal states are obligated by UNCLOS to "adopt laws and regulations to prevent, reduce and control pollution of the marine environment arising from or in connection with seabed activities subject to their jurisdiction", as well as "other measures that may be necessary to prevent, reduce, and control such pollution" (UNCLOS, Article 208 (1) and (2)), further specifying that "such laws, regulations and measures shall be no less effective than international rules, standards and recommended practices and procedures" (UNCLOS, Article 208 (3)). In this regard, UNCLOS, article 194 (3c) obligates states to minimize "pollution from installations and devices used in exploration or exploitation of the natural resources of the seabed and subsoil" (UNCLOS, Article 194 (3c)). This also includes the obligation of states to prevent transboundary harm arising from activities conducted in areas under their jurisdiction (UNCLOS, Article 194 (2)).

Several states have already enacted specific deep-sea mining regulations or incorporated them within existing frameworks. Papua New Guinea, has, for example, incorporated provisions for deep-sea mining in its 1992 Mining Act. The Mining Act aims mainly to encourage mining and contains very little environmental provisions. These are included in the 2000 Environment Act, which, for example, requires the submission of environmental impact statements (EIS) (including monitoring, environmental management programs, collection of baseline data and remediation), and Environmental Inception Reports (§51(b)). Past experience with terrestrial mining operations, as well as the country's high level of poverty, civil conflict, inequality and poor rule of law gives rise to concern, however, with respect to the implementation and enforcement of the regulations (Singh and Hunter 2019). Another Pacific island state interested in hosting deep-sea mining operations within their jurisdiction is Tonga, which has already issued exploration licenses to several contractors under the country's mineral and petroleum mining law (Blue Ocean Law and the Pacific Network on Globalisation 2016; Singh and Hunter 2019). In 2014, Tonga has, however, adopted its new Seabed Minerals Act, which has been drafted with the help of the Secretariat of the Pacific Community and the European Union. Although the Seabed

Minerals Act contains suitable environmental provisions, including the requirement to submit environmental impact assessments (EIA), it is doubtful that the country will be able to implement and enforce the regulations, due to a profound lack of financial and institutional capacity (Singh and Hunter 2019). The Cook Islands are actively seeking contractors to exploit nodules within its EEZ. The country adopted its Seabed Minerals Act in 2009, which mainly aimed at facilitating mining and gave little attention to environmental concerns. The 2015 Seabed Minerals (Protection and Exploration) Regulations contained more provisions on the environment, albeit in weak language. The country has, however, implemented the Marae Moana Act in 2017, which establishes the marine protected area Marae Moana, including a 50km no-mine zone around the country's coastline (§24). In contrast to the small island states, New Zealand, which incorporated provisions on deep-sea mining in its 1991 Crown Minerals Act and 2012 Exclusive Economic Zone and Continental Shelf (Environmental Effects) Act, appears to place greater emphasis on the protection of the environment and has even denied a mining application because of it (New Zealand EPA 2015; Singh and Hunter 2019).

4.3.2. The Area

The ISA has already issued three sets of prospecting and exploration regulations for nodules, crusts and SMS deposits and is currently in the process of developing a corresponding set of exploitation regulations. The draft application regulations contain requirements for the application for and approval of exploitation contracts, including the obligation to submit a plan of work, a mining plan, a feasibility report, a financing plan, a training plan, an emergency response and contingency plan, an environmental impact statement, an environmental management and monitoring plan, and a closure plan. The drafting process also included a stakeholder consultation phase, during which contractors identified gaps in the regulatory framework, including the lack of information on the operationalization of the polluter pays principle, the precautionary approach and the ecosystem approach, as well as the consideration of the impacts of climate change and cumulative effects. Furthermore, concerns were raised about the review of contractor compliance with environmental regulations and the unclarified relationship between environmental impact statements, environmental standards, and environmental management and monitoring plans. To this end, the contractors suggested the drafting of concrete guidelines for the preparation of environmental impact statements and environmental management, monitoring and closure plans, including the requirements for the collection of baseline data. The stakeholders, furthermore, called for the development of standards to ensure the protection of the marine environment (ISBA/26/C2). In addition to the exploration and exploitation guidelines, the ISA has issued the Recommendation for the Guidance of the Contractors for the Assessment of Possible Environmental Impacts Arising from Exploration for Marine Minerals in the Area (ISBA/19/LTC/8), which prescribes the collection of baseline data in the exploration areas employing best available technologies and to conduct environmental impact assessments before, during, and after the exploration activities. Although the recommendations are not legally binding, contractors are expected to follow them (Lodge 2015).

5. Economic Considerations

Whether deep-sea mining will yield net benefits and for whom, depends on numerous factors, including the occurrence, volume and composition of the mineral deposit to be mined, the capital and operational costs required for recovering them (especially in comparison to terrestrial mining), the development of the metal market, and whether the environmental costs of mining are considered (Jaeckel 2020; Folkersen et al. 2019; Mukhopadhyay et al. 2019; Van Nijen et al. 2019). Any predictions of the future profitability of deep-sea mining are complicated by persisting knowledge gaps, a high level of uncertainty, and the general difficulty of expressing environmental impacts in economic terms (Folkersen et al. 2019; Mukhopadhyay et al. 2019; Folkersen et al. 2018b). Where deep-sea mining is carried out in the Area, the profitability of deep-sea mining may also be influenced by the compensation of terrestrial-mining countries, which are negatively affected by metals obtained from deep-sea mining entering the global market, as demanded by the CHM (Christiansen et al. 2019). According to Van Nijen et al. (2019), this could likely occur with respect to the manganese market, which according to them is "shallow (low activity compared to the volume), non-transparent, and fragmented" (p. 579).

5.1. National Jurisdiction

Within national jurisdiction, states expect to benefit from hosting deep-sea mining operations in two ways: by receiving royalties from the contractors in exchange for the right to exploit the country's mineral resources, and by collecting corporate income tax (Mullins and Burns 2018). Particularly small island states appear to have high hopes to generate revenue for their economies by encouraging the development of a deep-sea mining industry. Although the economic benefits may be substantial given the countries low number of inhabitants, the income from deep-sea mining may in reality be limited, as royalties and tax rates will likely have to be set at a low level to incentivize mining (Mullins and Burns 2018; Cardno 2016). Furthermore, due to a lack of financial, technical and institutional capacity, the countries may undervalue the potential adverse environmental impacts associated with the exploitation of the resource, as well as any potential impacts on other economic sectors such as fishery and tourism (Christiansen et al. 2019). Moreover, asymmetric power relations, which occur when one partner is considerably stronger than the other and influences the terms of the contract in its favor, could further reduce the benefits for the host country. In the deep-sea mining context, this risk is particularly pronounced as many developing countries choose to enter into contracts with foreign mining companies and investors (Le Meur et al. 2018). This not only applies to areas within national jurisdiction but also to the Area, where several developing states act as sponsors for companies of their own nationality, but who are subsidiaries of large foreign corporations. Examples include Nauru Ocean Resources Inc., Tonga Offshore Mining Limited, and Marawa Research and Exploration Ltd., who are nationals of Nauru, Tonga, and Kiribati, respectively, but subsidiaries of the Canadian Company DeepGreen Minerals Inc.

If deep-sea mining is to take place, revenues generated by deep-sea mining will have to be carefully invested to ensure long-lasting benefits for the community. The development of an effective fiscal and revenue management framework prior to the commencement of mining is considered an essential pre-requisite in this regard (UNDP and UN Environment 2018). Such frameworks are recommended to include provisions on competitive procurement procedures, frequent independent audits of financial accounts, and the regular disclosure of non-commercial and non-confidential information to the public. Furthermore, transparency and the delineation of clear decision-making strategies are considered essential to minimize the risks of corruption and mismanagement of revenues (Sachs and Warner 1995; Ovesen et al. 2018).

An effective fiscal and revenue management regime can also limit the adverse impacts of asymmetric power relations (Le Meur et al. 2018). If managed poorly, the revenues obtained from mining may easily turn into a resource curse for the host countries, which has been frequently shown in the context of terrestrial mining. Particularly, developing countries which usually have less diversified economies, run the risk of becoming overly dependent on the extractive industry. In this case, countries become increasingly vulnerable to external economic shock caused by changes in commodity prices and production levels (Ovesen et al. 2018). Furthermore, they are prone to experience the Dutch disease, which describes a situation where economic growth in one sector, i.e., the extraction of a natural resource, leads to a decline in other sectors. The increased influx of foreign currencies as a consequence of the increased export of the resource may lead to the appreciation of the local currency, which may cause other sectors of the economy to become less competitive on the international market. The Dutch disease can be prevented or counteracted by developing clear budgetary plans, detailing in advance how and when revenues are to be invested in the short-, medium- and long-term (Soros 2007; Ovesen et al. 2018). Furthermore, the establishment of offshore wealth funds in foreign currencies outside the country has been identified as a measure to ensure economic security even after the revenues from deep-sea mining decline (Al-Hassan et al. 2013). If and how the Dutch disease may affect countries involved in deep-sea mining, has not yet been researched.

Particularly developing countries often lack the capacity to develop, implement and enforce effective legislative frameworks (Bradley and Swaddling 2018). This is critical, as structural and administrative weaknesses can lead to revenue losses and negatively affect the credibility of the framework among local and foreign investors (Ovesen et al. 2018). However, several organizations exist to assist governments with the development of fiscal and revenue management regimes, such as the Pacific Community (SPC) and the Pacific Financial Technical Assistance Center (PFTAC). The latter has, for example, aided the Cook Islands' Seabed Mineral Authority in developing a mining tax regime. Previously, the Commonwealth Secretariat Economic and Legal Section (ELS) had carried out a Seabed Minerals Fiscal Regime Analysis in 2012 and provided recommendations to the Cook Islands' government to consider in the preparation of its mining and fiscal regime to ensure consistency with international practice and stakeholder expectations. The Cook Islands' fiscal regime has recently been passed in parliament and will be administered by the islands' Ministry of Financial Economic Management (CI Seabed Minerals Authority 2019).

5.2. The Area

In the Area, the ISA is obligated by UNCLOS and the 1994 Agreement relating to the implementation of Part XI of UNCLOS (1994 IA) to develop a payment regime composed of a payment mechanism, which determines the financial contributions contractors have to make to the ISA in exchange for exploiting the resources of the Area (CHM), and a benefit-sharing mechanism, according to which the economic and non-economic benefits of deep-sea mining will be shared among all of the ISA's member states (UNCLOS, Article 140, (Van Nijen et al. 2019; Jaeckel 2020; Jaeckel et al. 2016). In developing the payment regime, the ISA has to follow six principles outlined in the 1994 IA, which demand that the payment mechanism must be "fair, non-discriminatory, simple, and within the range of payments prevailing for land-based mining" and contain a procedure for monitoring compliance (Jaeckel et al. 2016, p. 199). The process of the development of a payment mechanism is ongoing. Open question concern inter alia, the type and level of revenue raising charges to be contributed by the contractors and ways to account for the high risk of the contractors in developing emergent industry (Van Nijen et al. 2019). ISA consultants have suggested the implementation of a 2% ad valorem royalty during the early phase, which would later be increased to about 6% as the industry grows. In this case, about 70% of the proceedings would flow to the contractors, 2%–6% would be transferred to the ISA and the remainder would be paid as income tax to the country in which the contractor pays taxes (e.g., the sponsoring state) (The African Group 2018; Levin et al. 2020). The proposal by the ISA consultants has, however, been criticized by some of the ISA's member states, particularly by the African Group, which considers the revenue that would be raised by this scenario insufficient to compensate the ISA member states for the loss of resources in the Area (The African Group 2019; Levin et al. 2020).

Like the payment mechanism, the benefit-sharing mechanism is still being developed. However, neither UNCLOS nor the 1994 IA specify what the benefits of mankind entail and how they should be shared. This could, for example, include the direct re-distribution of the financial contributions from the contractors or the investment of their contributions into a fund (Christiansen et al. 2019). Given the current perspective on the level of royalties set by the ISA, it seems unlikely, however, that this will generate reasonable income for developing countries (The African Group 2018; Jaeckel 2020). The sharing of benefits could also include the provision of capacity-building opportunities and the sharing of scientific research findings. To this end, the ISA has, for example, initiated several training programs and issued several scholarships. Christiansen et al. (2019) point out that this could be improved through better organization and the establishment of "dedicated organs such as a school or university that systematically organizes education and capacity-building according to overarching educational goals" (p. 77). Furthermore, scientific data has, thus far, only been shared to a limited extent, although it has frequently been called for that particularly environmental data should be made available to the public (Seascape Consultants 2014; Jaeckel et al. 2016; ISBA/20/C/31 and ISBA/18/C/20).

6. Social Considerations

The potential social impacts of deep-sea mining have, thus far, received little attention in research. Their nature and magnitude, therefore, remain largely unknown.

Wherever deep-sea mining takes place in the vicinity of coastlines, concerns have been raised about potential direct and indirect impacts on fisheries and tourism (Koschinsky et al. 2018; Folkersen et al. 2018a; Roche and Bice 2013; Binney and Fleming 2016). In comparison to terrestrial mining operations, which often provide indirect employment opportunities through the development of settlements around mining operations, deep-sea mining will take place with little to no presence on land. Furthermore, deep-sea mining operations require highly skilled personnel with experience in the fields of offshore engineering, project management and shipboard services; it is, therefore, unlikely that many jobs will be filled by members of the local communities (Binney and Fleming 2016). Whether the inhabitants of coastal countries will benefit socially from deep-sea mining operation in their vicinity strongly depends on how their governments will choose to invest the revenues obtained from mining. If invested properly, the countries' additional income can contribute to the improvement of community and health services, infrastructure or affordable housing. Mismanagement and corruption, however, could negate any potentially positive impacts.

Whereas governments have generally responded positively to the prospects of hosting deep-sea mining operations in areas under their jurisdiction, local communities, as well as a number of national and international NGOs having assumed a more critical position (Koschinsky et al. 2018). This became particularly apparent in relation to the struggles of Nautilus Minerals, which are attributed in part to vehement community opposition. Although it has yet to be explored how people form their opinion of deep-sea mining (e.g., based on past experience with similar industries like terrestrial mining, on scientific facts or other factors), some insight could already be gained from the Nautilus Minerals case in Papua New Guinea. In relation to this project, Filer and Gabriel (2018) identified three different arguments frequently voiced by opponents to the Solwara 1 project. The first one emphasizes the application of the precautionary approach and, therefore, calls for an interruption of all mining-related activities until sufficient knowledge on its associated environmental impacts is available. The second argument is a religious or spiritual one, which portrays the ocean as a sacred space that must not be affected by mining. The third argument is of a legal nature and relates to the right of local communities of free, prior and informed (FPIC) consent, as stated in the United Nations Declaration on the Rights of Indigenous People. In the context of deep-sea mining, which will take place far offshore, it is, however, difficult to identify who would be entitled to FPIC (see Filer and Gabriel (2018) for a thorough assessment of this problem).

To increase the social sustainability of deep-sea mining operations, it is necessary to anticipate any potential social impacts prior to the commercialization of the activity. Important tools in this regard include social impact assessments (SIAs) (often included in EIAs) and the development of corresponding social impact management plans (SIMPs). Like their environmental counterparts, SIAs provide information about expected impacts to inform the decision-making of governments, stakeholders and the public, while SIMPs detail suitable response mechanisms. They further describe how potential positive impacts could be enhanced (Franks 2011; Franks and Vanclay 2013). Furthermore, more consideration should be given to FPIC and general stakeholder participation (see Singh and Hunter 2019 for an assessment of existing regulatory frameworks with respect to the incorporation of FPIC and stakeholder participation). Social impacts should, in any case, be a central component of deep-sea mining risk assessments.

7. Synthesis

7.1. Implications for Sustainable Development

Whether deep-sea mining can contribute to sustainability and sustainable development first and foremost depends on how sustainability is understood. In this regard, a distinction is commonly made between strong sustainability and weak sustainability. The concepts are closely linked to the five capitals theory, which assumes that there are different forms of capital: natural capital (e.g., natural resources, ecosystem services), financial capital (e.g., revenues), manufactured capital (e.g., goods, technology), human capital (e.g., work force, educational levels, skills of individuals), and social capital (e.g., norms, social networks, cooperation and trust) (Ang and van Passel 2012; Moldan et al. 2012). From a weak sustainability perspective, sustainability or sustainable development can be achieved by transforming one form of capital into another, as long as the overall stock of capital is maintained or increased. In contrast to this, proponents of the strong sustainability concept believe that the individual forms of capital need to be maintained in and of themselves. This is especially true for natural capital, as this is considered vital for the growth of the other forms of capital and, therefore, essentially irreplaceable by other forms of capital.

From a strong sustainability perspective, deep-sea mining would be inacceptable, as it not only describes the exploitation of a finite resource but will also be associated with substantial environmental impacts. From this perspective, the only viable option would be to reduce the demand for primary metals by increasing the rate of recycling, improving product design and increasing the longevity of products. This would also be in line with SDG 8, which calls for more sustainable consumption and production patterns. From a weak sustainability perspective, deep-sea mining could be considered sustainable if the conversion from natural capital (i.e., the resource in the ground and the in-tact ecosystem) into the other forms of capital (e.g., revenue, employment) would keep the overall level of capital constant. This requires a careful weighing of the benefits and costs of deep-sea mining.

By generating additional revenue for developing states through royalties and corporate income tax, deep-sea mining could theoretically contribute to achieving economic prosperity and human well-being, as, for example, called for by SDG 1 (ending poverty), SDG 2 (ending hunger), SDG 3 (health, well-being) and SDG 10 (reduce inequality within and among countries). Here, the CHM, which specifically requires the equitable sharing of the monetary and non-monetary benefits obtained from the exploitation of the marine mineral resources in the Area, is of particular importance (see also Christiansen et al. 2019). Furthermore, deep-sea mining can provide the metals required for producing the technology needed for the transition to a low-carbon economy. Crystalline photovoltaic panels, for example, contain substantial amounts of aluminum (Al), copper (Cu) and silver (Ag), as well as several other metals in smaller quantities. Wind turbines need significant quantities of iron (Fe), Cu, and Al. Electric vehicles typically use lithium-ion batteries to store electricity, which require metals like nickel (Ni), cobalt (Co), Al, and manganese (Mn) oxides, depending on the specific type of battery. In addition to this, electric vehicles and wind turbines often operate permanent magnet generators, which require significant quantities of rare earth elements (REEs), such as neodymium (Nd) and dysprosium (Dy) (Grandell et al. 2016). Many of these metals could likely eventually be extracted from marine mineral deposits. In this regard, deep-sea mining could contribute to achieving SDG 7 (sustainable and modern energy for all), specifically SDG 17.2 (by 2030, increase substantially the share of renewable energy in the global energy mix) and SDG 11 (make cities and human settlements, inclusive, safe, resilient and sustainable) if sustainable transport refers to electromobility (although this appears to be far-fetched). Following this line of reasoning, deep-sea mining could also indirectly contribute to achieving SDG 13 (take urgent action to combat climate change and its impacts).

However, deep-sea mining will entail the large scale and long-term destruction of the marine environment in and around the mine sites and cause inevitably the loss of biodiversity. In this regard, deep-sea mining stands in stark contrast to SDG 14 (sustainable life under water), specifically SDG 14.2 (sustainably manage and protect marine and coastal ecosystems to avoid significant adverse impacts, including by strengthening their resilience, and take action for their restoration in order to achieve healthy and productive oceans). The restoration of adversely affected deep-sea ecosystems is, however, particularly difficult and expensive. Furthermore, if ecosystem services, particularly the ability of the ocean and the seafloor to sequester carbon from the atmosphere, are compromised, deep-sea mining may also conflict with SDG 13 (combating climate change). Moreover, it is doubtful whether the revenues that could be generated by collecting royalties and income taxes (if paid to the host country), would be high enough to promote economic growth, improve social services and support institutions. Furthermore, mismanagement of revenues and the undervaluation of environmental impacts could cause the decline of other economic sectors, negatively affect the environment, and provoke social unrest. The latter may be the case particularly in developing countries which often lack the financial and institutional capacity to develop, implement and enforce sound regulatory frameworks.

7.2. Good Governance

If deep-sea mining cannot be prevented, it is important to reduce its adverse impacts as much as possible, for example, by implementing principles of good governance. The core characteristics of good governance include (1) rule of law, (2) accountability, (3) strategic vision, (4) responsiveness, (5) consensus orientation, (6) equity, and (7) effectiveness and efficiency (Kardos 2012). Although different institutions emphasize different elements, there is consensus that good governance is a crucial foundation of sustainable development. Ardron et al. (2018) have analyzed the role of transparency in the context of deep-sea mining in detail, which according to them, also relates to the elements of public participation and accountability. According to them, based on a thorough review of existing codes of conduct, regulations, international agreements, and voluntary standards, Ardron et al. (2018) identify six components of good practice in transparency and analyze to what extent the regulations and recommendations set forth by the ISA reflect these core aspects. They conclude that the ISA has been forward-thinking in some ways, for example, with respect to releasing information after a certain time period and the emphasis on the precautionary approach. Furthermore, they state that the draft exploitation regulations appear to indicate that transparency may be improving to a certain extent, for example, with respect to making exploitation contracts publicly accessible (although some have criticized that the ISA's effort is still not sufficient, see above). At the same time, the ISA's rules and regulations and procedures do not seem to reflect best practices. For instance, the application of the six components of transparency indicated weaknesses, such as the inaccessibility of annual reports, which are treated confidentially, unclear quality assurance, the lack of reporting on the compliance of states and contractors to ISA regulations, the lack of public participation as observers are not allowed to attend key committee meetings, and the limited possibility for civil society or state parties to request a review or appeal to decisions of the authority (Ardron et al. 2018).

Good governance also plays an important role, where developing countries are planning to host deep-sea mining operations in their EEZs or on their extended continental shelves. In these countries, the implementation and success of good governance principles is often limited by a lack of trained personnel capable of developing effective policy frameworks (e.g., fiscal and revenue management plans and environmental regulations), controlling the quality of impact assessments (e.g., EIAs and SIAs) and impact management plans (e.g., environmental management plans (EMPs), SIMPs), and monitoring compliance and enforcement. Capacity-building is, therefore, not only important with respect to minimizing the potential negative impacts of deep-sea mining, but also with respect to maximizing potential benefits of the activity. The Natural Resource Charter also provides guidance for "governments, societies and the international community", although their implementation may be challenging (Cust and Manley 2014, page 4).

Kung et al. (2020) highlight that "uncertainties are translating into defects in emergent [deep-sea mining] governance architecture", both within and beyond the limits of national jurisdiction (p. 8). They highlight in particular, that applying EIA methodology, albeit a well-established process, is difficult in the context of deep-sea mining, which is "a frontier industry with scant environmental data on the status quo, and with no functional precedent in in terms of project design" (ibid., p. 9). In contrast to terrestrial activities, which usually benefit from information of experiences made with similar processes in similar environmental settings, there is no such option for deep-sea mining. Furthermore, there is no definition yet of what actually constitutes serious harm. Experience from terrestrial mining can, however, be used, where conflicts of ownership or between users of the marine environment occur.

Independent of the decision for or against deep-sea mining, research on deep-sea ecosystems and potential environmental, economic and social impacts of deep-sea mining should be continued, as the past decades have shown that the interest in deep-sea mineral deposits may periodically reoccur and future generations should have a solid foundation of knowledge to make decisions based on scientific facts.

Author Contributions: Conceptualization, L.H. and A.K.; writing—original draft preparation, L.H.; writing—review and editing, L.H. and A.K.; visualization, L.H. and A.K.; supervision, A.K. All authors have read and agreed to the published version of the manuscript.

Funding: This research received no external funding.

Conflicts of Interest: The authors declare no conflict of interest.

References

Aldred, Jessica. 2019. The Future of Deep Seabed Mining. Available online: https://chinadialogueocean.net/6682-future-deep-seabed-mining/ (accessed on 18 November 2020).

Al-Hassan, Abdullah, Michael G. Papaioannou, Martin Skancke, and Cheng Chih Sung. 2013. *Sovereign Wealth Funds: Aspects of Governance Structures and Investment Management.* Washington, DC: International Monetary Fund.

Amon, Diva J., Amanda F. Ziegler, Thomas G. Dahlgren, Adrian G. Glover, Aurélie Goineau, Andrew J. Gooday, Helena Wiklund, and Craig R. Smith. 2016. Insights into the Abundance and Diversity of Abyssal Megafauna in a Polymetallic-Nodule Region in the Eastern Clarion-Clipperton Zone. *Scientific Reports* 6: 30492. [CrossRef] [PubMed]

Ang, Frederic, and Steven van Passel. 2012. Beyond the Environmentalist's Paradox and the Debate on Weak versus Strong Sustainability. *BioScience* 62: 251–59. [CrossRef]

Ardron, Jeff A., Henry A. Ruhl, and Daniel O. B. Jones. 2018. Incorporating Transparency into the Governance of Deep-Seabed Mining in the Area beyond National Jurisdiction. *Marine Policy* 89: 58–66. [CrossRef]

Armstrong, Claire W., Naomi S. Foley, Rob Tinch, and Sybille van den Hove. 2012. Services from the Deep: Steps towards Valuation of Deep Sea Goods and Services. *Ecosystem Services* 2: 2–13. [CrossRef]

Atmanand, Malayath A., and Gidugu A. Ramadass. 2017. Concepts of Deep- Sea Mining Technologies. In *Deep-Sea Mining Resource Potential, Technical and Environmental Considerations.* Edited by Rahul Sharma. Cham: Springer International Publishing, pp. 305–43. [CrossRef]

Balvanera, Patricia, Ilyas Siddique, Laura Dee, Alain Paquette, Forest Isbell, Andrew Gonzalez, Jarrett Byrnes, Mary I. O'Connor, Bruce A. Hungate, and John N. Griffin. 2014. Linking Biodiversity and Ecosystem Services: Current Uncertainties and the Necessary next Steps. *BioScience* 64: 49–57. [CrossRef]

Batker, David, and Rowan Schmidt. 2015. *Environmental and Social benchmarking Analysis of Nautilus Minerals Inc. Solwara I Project.* Tacoma: Earth Economics, Available online: https://www.eartheconomics.org/publications-archive (accessed on 17 October 2019).

Bennett, Elena M., Wolfgang Cramer, Alpina Begossi, Georgina Cundill, Sandra Díaz, Benis N. Egoh, Ilse R. Geijzendorffer, Cornelia B. Krug, Sandra Lavorel, Elena Lazos, and et al. 2015. Linking Biodiversity, Ecosystem Services, and Human Well-Being: Three Challenges for Designing Research for Sustainability. *Current Opinion in Environmental Sustainability* 14: 76–85. [CrossRef]

Billet, David, Daniel O. B. Jones, and Philip P. E. Weaver. 2019. Improving Environmental Management in Deep-Sea Mining. In *Environmental Issues of Deep-Sea Mining: Impacts, Consequences and Policy Perspectives*. Edited by Rahul Sharma. Cham: Springer International Publishing, pp. 403–46.

Binney, Jim, and Chris Fleming. 2016. *Counting the Potential Cost of Deep Sea-Bed Mining to Fiji: A Report for WWF International*. Toowong: MainStream Economics and Policy, Available online: https://wwfint.awsassets.panda.org/downloads/deep_seabed_mining___economic_risks___final_2.pdf (accessed on 10 December 2020).

Blue Mining. 2014. Blue Mining: Breakthrough Solutions for Sustainable Deep-Sea Mining. Available online: https://bluemining.eu/facts-and-figures/ (accessed on 7 November 2019).

Blue Ocean Law and the Pacific Network on Globalisation. 2016. *Resource Roulette: How Deep-Sea Mining and Inadequate Regulatory Frameworks Imperil the Pacific and Its Peoples*. Available online: http://www.savethehighseas.org/wp-content/uploads/2018/05/Blue-oceans-law-Resource_Roulette.pdf (accessed on 9 December 2020).

Bundesministerium für Umwelt Naturschutz und nukleare Sicherheit (BMU). 2015. Die 2030-Agenda Für Nachhaltige Entwicklung. Available online: https://www.bmu.de/themen/europa-internationales-nachhaltigkeit-digitalisierung/nachhaltige-entwicklung/2030-agenda/ (accessed on 17 October 2020).

Boschen, Rachel E., Ashley A. Rowden, Malcolm R. Clark, Arne Pallentin, and Jonathan P. A. Gardner. 2016. Seafloor Massive Sulfide Deposits Support Unique Megafaunal Assemblages: Implications for Seabed Mining and Conservation. *Marine Environmental Research* 115: 78–88. [CrossRef]

Bourrel, Marie, Torsten Thiele, and Duncan Currie. 2018. The Common of Heritage of Mankind as a Means to Assess and Advance Equity in Deep Sea Mining. *Marine Policy* 95: 311–16. [CrossRef]

Bradley, Melanie, and Alison Swaddling. 2018. Addressing Environmental Impact Assessment Challenges in Pacific Island Countries for Effective Management of Deep Sea Minerals. *Marine Policy* 95: 356–62. [CrossRef]

Brundtland, Gro Harlem. 1987. Report of the World Commission on Environment and Development: Our Common Future. A/42/427. United Nations. Available online: https://digitallibrary.un.org/record/139811 (accessed on 15 October 2019).

Calvo, Guiomar, Gavin Mudd, Alicia Valero, and Antonio Valero. 2016. Decreasing Ore Grades in Global Metallic Mining: A Theoretical Issue or a Global Reality? *Resources* 5: 36. [CrossRef]

Cardno. 2016. *An Assessment of the Costs and Benefits of Mining Deep-Sea Minerals in the Pacific Island Region*. Suva: Pacific Community (SPC), Available online: https://www.sprep.org/attachments/VirLib/Regional/deep-sea-mining-cba-PICs-2016.pdf (accessed on 17 October 2019).

Christiansen, Sabine, Duncan Currie, Kate Houghton, Alexander Müller, Manuel Rivera, Oscar Schmidt, Prue Taylor, and Sebastian Unger. 2019. *Towards a Contemporary Vision for the Global Seafloor—Implementing the Common Heritage of Mankind*. Berlin: Heinrich Boell Stiftung, vol. 45, Available online: https://www.boell.de/en/2019/11/11/towards-contemporary-vision-global-seafloor-implementing-common-heritage-mankind (accessed on 19 November 2020).

CI Seabed Minerals Authority. 2019. Priorities: Development and Growth. Available online: https://www.seabedmineralsauthority.gov.ck/priorities (accessed on 14 November 2019).

Clark, Malcom R., Ashley A. Rowden, Thomas Schlacher, Alan Williams, Mireille Consalvey, Karen I. Stocks, Alex D. Rogers, Timothy D. O'Hara, Martin White, Timothy M. Shank, and et al. 2010. The Ecology of Seamounts: Structure, Function, and Human Impacts. *Annual Review of Marine Science* 2: 253–78. [CrossRef]

Copley, Jon T., Leigh Marsh, Adrian G. Glover, Veit Hühnerbach, Verity E. Nye, William D. K. Reid, Christopher J. Sweeting, Ben D. Wigham, and Helena Wiklund. 2016. Ecology and Biogeography of Megafauna and Macrofauna at the First Known Deep-Sea Hydrothermal Vents on the Ultraslow-Spreading Southwest Indian Ridge. *Scientific Reports* 6: 39158. [CrossRef] [PubMed]

Cust, Jim, and David Manley. 2014. *Natural Resource Charter*, 2nd ed. London: Natural Resource Institute, Available online: https://resourcegovernance.org/sites/default/files/NRCJ1193_natural_resource_charter_19.6.14.pdf (accessed on 10 October 2020).

Cuvelier, Daphne, Sabine Gollner, Daniel O. B. Jones, Stefanie Kaiser, Pedro Martínez Arbizu, Lena Menzel, Nélia C. Mestre, Telmo Morato, Christopher Pham, Florence Pradillon, and et al. 2018. Potential Mitigation and Restoration Actions in Ecosystems Impacted by Seabed Mining. *Frontiers in Marine Science* 5: 1–22. [CrossRef]

Drazen, Jeffrey C., Craig R. Smith, Kristina M. Gjerde, Steven H. D. Haddock, Glenn S. Carter, C. Anela Choy, Malcolm R. Clark, Pierre Dutrieux, Erica Goetze, Chris Hauton, and et al. 2020. Midwater Ecosystems Must Be Considered When Evaluating Environmental Risks of Deep-Sea Mining. *Proceedings of the National Academy of Sciences of the United States of America* 117: 17455–60. [CrossRef]

Filer, Colin, and Jennifer Gabriel. 2018. How Could Nautilus Minerals Get a Social Licence to Operate the World's First Deep Sea Mine? *Marine Policy* 95: 394–400. [CrossRef]

Folkersen, Maja Vinde, Christopher M. Fleming, and Syezlin Hasan. 2018a. Deep Sea Mining's Future Effects on Fiji's Tourism Industry: A Contingent Behaviour Study. *Marine Policy* 96: 81–89. [CrossRef]

Folkersen, Maja Vinde, Christopher M. Fleming, and Syezlin Hasan. 2018b. The Economic Value of the Deep Sea: A Systematic Review and Meta-Analysis. *Marine Policy* 94: 71–80. [CrossRef]

Folkersen, Maja Vinde, Christopher M. Fleming, and Syezlin Hasan. 2019. Depths of Uncertainty for Deep-Sea Policy and Legislation. *Global Environmental Change* 54: 1–5. [CrossRef]

Frakes, Jennifer. 2003. The Common Heritage of Mankind Principle and the Deep Seabed, Outer Space, and Antarctica: Will Developed and Developing Nations Reach a Compromise. *Wisconsin International Law Journal* 21: 409–34.

Franks, Daniel M., and Frank Vanclay. 2013. Social Impact Management Plans: Innovation in Corporate and Public Policy. *Environmental Impact Assessment Review* 43: 40–48. [CrossRef]

Franks, Daniel M. 2011. Management of the Social Impacts of Mining. In *SME Mining Engineering Handbook*, 1st ed. Edited by P. Darling. Littleton: Society for Mining, Metallurgy, and Exploration, pp. 1817–25.

German, Christopher R., Sven Petersen, and Mark D. Hannington. 2016. Hydrothermal Exploration of Mid-Ocean Ridges: Where Might the Largest Sulfide Deposits Be Forming? *Chemical Geology* 420: 114–26. [CrossRef]

Gillard, Ben, Kaveh Purkiani, Damianos Chatzievangelou, and Annemiek Vink. 2019. Physical and Hydrodynamic Properties of Deep Sea Mining-Generated, Abyssal Sediment Plumes in the Clarion Clipperton Fracture Zone (Eastern-Central Pacific). *Elementa Science of the Anthropocene* 7: 5. [CrossRef]

Grandell, Leena, Antti Lehtil, Mari Kivinen, Tiina Koljonen, Susanna Kihlman, and Laura S. Lauri. 2016. Role of Critical Metals in the Future Markets of Clean Energy Technologies. *Renewable Energy* 95: 53–62. [CrossRef]

Haeckel, Matthias, Iris König, Volkher Riech, Michael E. Weber, and Erwin Suess. 2001. Pore Water Profiles and Numerical Modelling of Biogeochemical Processes in Peru Basin Deep-Sea Sediments. *Deep Sea Research Part II: Topical Studies in Oceanography* 48: 3713–36. [CrossRef]

Halbach, Peter, Frank T. Manheim, and Peter Otten. 1982. Co-Rich Ferrmanganese Deposits in the Marginal Seamount Regions of the Central Pacific Basin—Results of the Midpac'81. *Erzmetall* 35: 447–53.

Hannington, Mark D., Cornel E. J. De Ronde, and Sven Petersen. 2005. *Sea-Floor Tectonics and Submarine Hydrothermal Systems*. Economic Geology. Littelton: Society of Economic Geologists, pp. 111–41.

Hannington, Mark D., John Jamieson, Thomas Monecke, Sven Petersen, and Stace Beaulieu. 2011. The Abundance of Seafloor Massive Sulfide Deposits. *Geology* 39: 1155–58. [CrossRef]

Hauton, Chris, Alastair Brown, Sven Thatje, Nélia C. Mestre, Maria J. Bebianno, Inês Martins, Raul Bettencourt, Miquel Canals, Anna Sanchez-Vidal, Bruce Shillito, and et al. 2017. Identifying Toxic Impacts of Metals Potentially Released during Deep- Sea Mining—A Synthesis of the Challenges to Quantifying Risk. *Frontiers in Marine Science* 4: 1–13. [CrossRef]

Hein, James R., and Andrea Koschinsky. 2014. Deep-Ocean Ferromanganese Crusts and Nodules. *Treatise on Geochemistry* 13: 273–91.

Hein, James R., Marjorie S. Schulz, and Lisa M. Gein. 1992. Central Pacific Cobalt-Rich Ferromanganese Crusts: Historical Perspective and Regional Variability. In *Geology and Offshore Mineral Resources of the Central Pacific Basin*. Edited by B. H. Keating and B. R. Bolton. New York: Springer, pp. 261–83.

Hein, James R., Kira Mizell, Andrea Koschinsky, and Tracey A. Conrad. 2013. Deep-Ocean Mineral Deposits as a Source of Critical Metals for High- and Green-Technology Applications: Comparison with Land-Based Resources. *Ore Geology Reviews* 51: 1–14. [CrossRef]

Heinrich, Luise, Andrea Koschinsky, Till Markus, and Pradeep Singh. 2020. Quantifying the Fuel Consumption, Greenhouse Gas Emissions and Air Pollution of a Potential Commercial Manganese Nodule Mining Operation. *Marine Policy* 114: 103678. [CrossRef]

Hong, Sup, Hyung-Woo Kim, Jong-su Choi, Tae-Kyeong Yeu, Soung-Jae Park, Chang-Ho Lee, and Suk-Min Yoon. 2010. A Self-Propelled Deep-Seabed Miner and Lessons from Shallow Water Tests. Paper presented at the ASME 29th International Conference on Ocean, Offshore and Arctic Engineering, Shanghai, China, June 6–11; New York: ASME, vol. 5, pp. 75–86.

Huijbregts, Mark A. J., Zoran J. N. Steinmann, Pieter M. F. Elshout, Gea Stam, Francesca Verones, Marisa Vieira, Anne Hollander, Michiel Zijp, and Rosalie van Zelm. 2016. *ReCiPe: A Harmonized Life Cycle Impact Assessment Method at Midpoint and Endpoint Level Report 1: Characterization*. Bilthoven: National Institute for Public Health and the Environment.

International Maritime Organization (IMO). 2015. *Third IMO Greenhouse Gas Study 2014*. London: International Maritime Organization, Available online: https://www.imo.org/en/OurWork/ Environment/Pages/Greenhouse-Gas-Studies-2014.aspx (accessed on 6 October 2020).

International Seabed Authority (ISA). 2020. Exploration Contracts. Available online: https: //isa.org.jm/exploration-contracts (accessed on 12 December 2020).

International Tribunal for the Law of the Sea (ITLOS). 2011. *Responsibilities and Obligations of States with Respect to Activities in the Area, Advisory Opinion*. ITLOS Reports. Hamburg: ITLOS.

Jacob, Céline, Sylvain Pioch, and Sébastien Thorin. 2016. The Effectiveness of the Mitigation Hierarchy in Environmental Impact Studies on Marine Ecosystems: A Case Study in France. *Environmental Impact Assessment Review* 60: 83–98. [CrossRef]

Jaeckel, Aline, Jeff A. Ardron, and Kristina M. Gjerde. 2016. Sharing Benefits of the Common Heritage of Mankind—Is the Deep Seabed Mining Regime Ready? *Marine Policy* 70: 198–204. [CrossRef]

Jaeckel, Aline. 2020. Benefitting from the Common Heritage of Humankind: From Expectation to Reality. *The International Journal of Marine and Coastal Law* 35: 1–22. [CrossRef]

Jankowski, Jacek A., and Werner Zielke. 2001. The Mesoscale Sediment Transport Due to Technical Activities in the Deep Sea. *Deep-Sea Research Part II: Topical Studies in Oceanography* 48: 3487–521. [CrossRef]

Japan Oil Gas and Metals National Corporation (JOGMEC). 2020. JOGMEC Conducts World's First Successful Excavation of Cobalt-Rich Seabed in the Deep Ocean. Available online: http://www.jogmec.go.jp/english/news/release/news_01_000033.html (accessed on 29 November 2020).

Jones, Daniel O. B., Jennifer M. Durden, Kevin Murphy, Kristina M. Gjerde, Aleksandra Gebicka, Ana Colaço, Telmo Morato, Daphne Cuvelier, and David S. M. Billet. 2019. Existing environmental management approaches relevant to deep-sea mining. *Marine Policy* 103: 172–81. [CrossRef]

Joyner, Christopher C. 1986. Legal implications of the concept of the common heritage of mankind. *International and Comparative Law Quarterly* 35: 190–99. [CrossRef]

Kardos, Mihaela. 2012. The reflection of good governance in sustainable development strategies. *Procedia - Social and Behavioral Sciences* 58: 1166–73. [CrossRef]

Kiss, Alexandre. 1985. The Common Heritage of Mankind: Utopia or Reality? *Law in the International Community* 40: 423–41. [CrossRef]

Koschinsky, Andrea, Christian Borowski, and Peter Halbach. 2003. Reactions of the Heavy Metal Cycle to Industrial Activities in the Deep Sea: An Ecological Assessment. *International Review of Hydrobiology* 88: 102–27. [CrossRef]

Koschinsky, Andrea, Luise Heinrich, Klaus Boehnke, J. Christopher Cohrs, Till Markus, Maor Shani, Pradeep Singh, Karen Smith Stegen, and Welf Werner. 2018. Deep-Sea Mining: Interdisciplinary Research on Potential Environmental, Legal, Economic, and Societal Implications. *Integrated Environmental Assessment and Management* 14: 672–91. [CrossRef]

Koschinsky, Andrea. 2016. Sources and Forms of Trace Metals Taken up by Hyrothermal Vent Mussels, and Possible Adaption and Mitigation Strategies. In *Trace Metal Biogeochemistry and Ecology of Deep Sea Hydrothermal Vent Systems*, 50th ed. Edited by L. Demina and S. Galkin. Cham: Springer International Publishing. [CrossRef]

Kung, Anthony, Kamila Svobodova, Eléonore Lèbre, Rick Valenta, Deanna Kemp, and John R. Owen. 2020. Governing Deep Sea Mining in the Face of Uncertainty. *Journal of Environmental Management* 279: 111593. [CrossRef]

Le, Jennifer T., Lisa A. Levin, and Richard T. Carson. 2017. Incorporating Ecosystem Services into Environmental Management of Deep-Seabed Mining. *Deep-Sea Research Part II: Topical Studies in Oceanography* 137: 486–503. [CrossRef]

Le Meur, Pierre-Yves, Nicholas Arndt, Patrice Christmann, and Vincent Geronimi. 2018. Deep-Sea Mining Prospects in French Polynesia: Governance and the Politics of Time. *Marine Policy* 95: 380–87. [CrossRef]

Levin, Lisa A., Diva J. Amon, and Hannah Lily. 2020. Challenges to the Sustainability of Deep-Seabed Mining. *Nature Sustainability* 3: 784–94. [CrossRef]

Lily, Hannah. 2018. Sponsoring State Approaches to Liability Regimes for Environmental Damage Caused by Seabed Mining. *Liability Issues for Deep Seabed Mining Series* 3: 13.

Lodge, Michael. 2015. The Deep Seabed. In *The Oxford Handbook of the Law of the Sea*. Edited by Donald Rothwell, Alex Oude Efferink, Karen Scott and Tim Stephens. Oxford: Oxford University Press.

Lusty, Paul A. J., James R. Hein, and Pierre Josso. 2018. Formation and Occurrence of Ferromanganese Crusts: Earth's Storehouse for Critical Metals. *Elements* 14: 313–18. [CrossRef]

Millennium Ecosystem Assessment (MEA). 2005. *Ecosystems and Human Well-Being: Synthesis*. Washington, DC: Island Press, vol. 5.

Mero, John L. 1965. *The Mineral Resources of the Sea*. Amsterdam: Elsevier Publishing Co.

Mewes, Konstantin, José M. Mogollón, Aude Picard, Carsten Rühlemann, Thomas Kuhn, Kerstin Nöthen, and Sabine Kasten. 2014. Impact of Depositional and Biogeochemical Processes on Small Scale Variations in Nodule Abundance in the Clarion-Clipperton Fracture Zone. *Deep-Sea Research Part I: Oceanographic Research Papers* 91: 125–41. [CrossRef]

Moldan, Bedřich, Svatava Janoušková, and Tomáš Hák. 2012. How to Understand and Measure Environmental Sustainability: Indicators and Targets. *Ecological Indicators* 17: 4–13. [CrossRef]

Monecke, Thomas, Sven Petersen, and Mark D. Hannington. 2014. Constraints on Water Depth of Massive Sulfide Formation: Evidence from Modern Seafloor Hydrothermal Systems in Arc- Related Settings. *Economic Geology* 109: 2079–101. [CrossRef]

Mukhopadhyay, Ranadhir, Sankalp Naik, Shawn De Souza, Ozinta Dias, Sridhar D. Iyer, and Anil K. Ghosh. 2019. The Economics of Mining Seabed Manganese Nodules: A Case Study of the Indian Ocean Nodule Field. *Marine Georesources and Geotechnology* 37: 845–51. [CrossRef]

Mullingneaux, Lauren S. 1987. Organisms Living on Manganese Nodules and Crusts: Distribution and Abundance at Three North Pacific Sites. *Deep Sea Research* 2: 165–84. [CrossRef]

Mullins, Peter, and Lee Burns. 2018. The Fiscal Regime for Deep Sea Mining in the Pacific Region. *Marine Policy* 95: 337–45. [CrossRef]

New Zealand Environmental Protection Authority. 2015. Decision on Marine Consent Application Chatham Rock Phoshpate Limited to Mine Phosphorite Nodules on the Chatham Rise. Available online: https://cer.org.za/wp-content/uploads/2016/08/EPA-New-Zealand-Chatham-Rock-Phosphate-Decision.pdf (accessed on 9 December 2020).

Niner, Holly J., Jeff A. Ardron, Elva G. Escobar, Matthew Gianni, Aline Jaeckel, Daniel O. B. Jones, Lisa A. Levin, Craig R. Smith, Torsten Thiele, Phillip J. Turner, and et al. 2018. Deep-Sea Mining with No Net Loss of Biodiversity-an Impossible Aim. *Frontiers in Marine Science* 5: 53. [CrossRef]

Orcutt, Beth N., James A. Bradley, William J. Brazelton, Emily R. Estes, Jacqueline M. Goordial, Julie A. Huber, Rose M. Jones, Nagissa Mahmoudi, Jeffrey J. Marlow, Sheryl Murdoch, and et al. 2020. Impacts of Deep-Sea Mining on Microbial Ecosystem Services. *Limnology and Oceanography* 65: 1489–510. [CrossRef]

Ovesen, Vidar, Ron Hackett, Lee Burns, Peter Mullins, and Scott Roger. 2018. Managing Deep Sea Mining Revenues for the Public Good—Ensuring Transparency and Distribution Equity. *Marine Policy* 95: 332–36. [CrossRef]

Paul, Sophie A. L., Birgit Gaye, Matthias Haeckel, Sabine Kasten, and Andrea Koschinsky. 2018. Biogeochemical Regeneration of a Nodule Mining Disturbance Site: Trace Metals, DOC and Amino Acids in Deep-Sea Sediments and Pore Waters. *Frontiers in Marine Science* 5: 117. [CrossRef]

Paulinkas, Daina, Steven Katona, Erika Ilves, Greg Stone, and Anthony O'Sullivan. 2020. *Where Should Metals for the Green Transition Come from? Comparing Environmental, Social, and Economic Impacts of Supplying Base Metals from Land Ores and Seafloor Polymetallic Nodules.* Vancouver: DeepGreen, Available online: https://deep.green/wp-content/uploads/2020/04/LCA-White-Paper_Where-Should-Metals-for-the-Green-Transition-Come-From_FINAL_low-res.pdf (accessed on 15 November 2020).

Petersen, Sven, Anna Krätschell, Nico Augustin, John Jamieson, James R. Hein, and Mark D. Hannington. 2016. News from the Seabed-Geological Characteristics and Resource Potential of Deep- Sea Mineral Resources. *Marine Policy* 70: 175–87. [CrossRef]

Popper, Arthur R., Jane Fewtrell, Michael E. Smith, and Robert D. McCauley. 2003. Anthropogenic Sound: Effects on the Behavior and Physiology of Fishes. *Marine Technology Society Journal* 37: 35–40. [CrossRef]

Ramboll IMS & HWWI. 2016. *Analyse Des Volkswirtschaftlichen Nutzens Der Entwicklung Eines Kommerziellen Tiefseebergbaus in Den Gebieten, in Denen Deutschland Explorationslizenzen Der Internationalen Meeresbodenbehörde Besitzt.* Hamburg: Ramboll IMS Ingenieursgesellschaft mbH, Available online: https://www.bmwi.de/Redaktion/DE/Publikationen/Studien/analyse-des-volkswirtschaftlichen-nutzens-der-entwicklung-eines-kommerziellen-tiefseebergbaus.pdf?__blob=publicationFile&v=6 (accessed on 17 October 2019).

Roche, Charles, and Sarah Bice. 2013. Anticipating Social and Community Impacts of Deep Sea Mining. *Deep Sea Minerals and the Green Economy*, 59–80.

Rogers, Alex D., Paul A. Tyler, Douglas P. Connelly, Jon T. Copley, Rachael James, Robert D. Larter, Katrin Linse, Rachel A. Mills, Alfredo Naveira Garabato, Richard D. Pancost, and et al. 2012. The Discovery of New Deep-Sea Hydrothermal Vent Communities in the Southern Ocean and Implications for Biogeography. *PLoS Biology* 10: e1001234. [CrossRef]

Rolinski, Susanne, Joachim Segschneider, and Jürgen Sündermann. 2001. Long-Term Propagation of Tailings from Deep-Sea Mining under Variable Conditions by Means of Numerical Simulations. *Deep-Sea Research Part II: Topical Studies in Oceanography* 48: 3469–85. [CrossRef]

Rowden, Ashley A., Thomas A. Schlacher, Alan Williams, Malcolm R. Clark, Robert Stewart, Franziska Althaus, David A. Bowden, Mireille Consalvey, Wayne Robinson, and Joanne Dowdney. 2010. A Test of the Seamount Oasis Hypothesis: Seamounts Support Higher Epibenthic Megafaunal Biomass than Adjacent Slopes. *Marine Ecology* 31: 95–106. [CrossRef]

Rühlemann, Carsten, Thomas Kuhn, and Annemiek Vink. 2019. Tiefseebergbau—Ökologische Und Sozioökonomische Auswirkungen. *Bürger & Staat* 69: 226–36. Available online: https://www.buergerundstaat.de/4_19/ozean_meere.pdf#page=40 (accessed on 19 November 2020).

Sachs, Jeffrey D., and Andrew M. Warner. 1995. *Natural Resource Abundance and Economic Growth*. Cambridge, MA: NBER Working Paper Series, Available online: https://www.nber.org/system/files/working_papers/w5398/w5398.pdf (accessed on 17 October 2019).

Seascape Consultants. 2014. Review of Implementation of the Environmental Management Plan for the Clarion Clipperton Zone. Available online: https://isa.org.jm/files/documents/EN/20Sess/LTC/CCZ-EMPRev.pdf (accessed on 5 November 2019).

Singh, Pradeep, and Julie Hunter. 2019. Protection of the Marine Environment: The International and National Regulation of Deep Seabed Mining Activities. In *Environmental Issues of Deep-Sea Mining: Impacts, Consequences and Policy Perspectives*. Edited by Rahul Sharma. Cham: Springer International Publishing, pp. 471–503.

Soros, George. 2007. *Escaping the Resource Curse*. Edited by H. Marcatan, J. D. Sachs and J. E. Stiglitz. New York: Columbia University Press.

Spagnoli, Giovanni, Johann Rongau, Julien Denegre, Stape A. Miedema, and Leonhard Weixler. 2016. A Novel Mining Approach for Seafloor Massive Sulfide Deposits. Paper presented at Offshore Technology Conference, Houston, TX, USA, May 2–5; Houston: Offshore Technology Conference.

Sparenberg, Ole. 2019. A Historical Perspective on Deep-Sea Mining for Manganese Nodules, 1965–2019. *The Extractive Industries and Society* 6: 842–54. [CrossRef]

SPC. 2013a. *Deep Sea Minerals: Cobalt-Rich Ferromanganese Crusts, a Physical, Biological, Environmental, and Technical Review*. Edited by E. Baker and Y. Beaudoin. Noumea: Secretariat of the Pacific Community, vol. 1C, Available online: https://cld.bz/bookdata/Da9poHo/basic-html/index.html#1 (accessed on 15 October 2019).

SPC. 2013b. *Deep Sea Minerals: Manganese Nodules—A Physical, Biological, Environmental, and Technical Review*. Edited by E. Baker and Y. Beaudoin. Noumea: Secretariat of the Pacific Community, vol. 1B, Available online: https://cld.bz/bookdata/h1Tu26r/basic-html/index.html#1 (accessed on 15 October 2019).

SPC. 2013c. *Deep Sea Minerals: Sea-Floor Massive Sulphides—A Physical, Biological, Environmental, and Technical Review*. Edited by E. Baker and Y. Beaudoin. Noumea: Secretariat of the Pacific Community, vol. 1A, Available online: https://cld.bz/bookdata/iHB2DZo/basic-html/index.html#1 (accessed on 15 October 2019).

Swaddling, Alison. 2016. *Pacific-ACP States Regional Environmental Management Framework for Deep-Sea Minerals Exploration and Exploitation*. Suva: Pacific Community.

The African Group. 2018. Statement by the Permanent Mission of Algeria to the International Seabed Authority. Available online: https://isa.org.jm/files/files/documents/alg-oboag-entp.pdf (accessed on 11 December 2020).

The African Group. 2019. Statement by the Permanent Mission of Algeria to the International Seabed Authority. Available online: https://www.isa.org.jm/files/files/documents/1-algeriaoboag_finmodel.pdf (accessed on 11 December 2020).

Thornborough, Kate J., S. Kim Juniper, S. Smith, and Lynn-Wei Wong. 2019. Towards an Ecosystem Approach to Environmental Impact Assessment for Deep-Sea Mining. In *Environmental Issues of Deep-Sea Mining: Impacts, Consequences and Policy Perspectives*. Edited by Rahul Sharma. Cham: Springer International Publishing, pp. 63–94.

Tunnicliffe, Verena, Anna Metaxas, Jennifer Le, Eva Ramirez-Llodra, and Lisa A. Levin. 2020. Strategic Environmental Goals and Objectives: Setting the Basis for Environmental Regulation of Deep Seabed Mining. *Marine Policy* 114: 103347. [CrossRef]

UNDP and UN Environment. 2018. *Managing Mining for Sustainable Development: A Sourcebook*. Bangkok: UNDP Bangkok Regional Hub and Poverty-Initiative Asia-Pacific of UNDP and UN Environment, Available online: https://www.undp.org/content/undp/en/home/librarypage/poverty-reduction/Managing-Mining-for-SD.html (accessed on 2 October 2019).

Van Dover, Cindy L., James Aronson, Linwood Pendleton, Samantha Smith, Sophie Arnaud-Haond, David Moreno-Mateos, Edward Barbier, David Billet, Keith Bowers, Roberto Danovaro, and et al. 2014. Ecological Restoration in the Deep Sea: Desiderata. *Marine Policy* 44: 98–106. [CrossRef]

Van Dover, Cindy L., Sophie Arnaud-Haond, Matthew Gianni, Stefan Helmreich, Julie A. Huber, Aline L. Jaeckel, Anna Metaxas, Linwood H. Pendleton, Sven Petersen, Eva Ramirez-Llodra, and et al. 2018. Scientific Rationale and International Obligations for Protection of Active Hydrothermal Vent Ecosystems from Deep-Sea Mining. *Marine Policy* 90: 20–28. [CrossRef]

Van Dover, Cindy L., Ana Colaço, Patrick C. Collins, Peter Croot, Anna Metaxas, Bramley J. Murton, Alison Swaddling, Rachel E. Boschen-Rose, Jens Carlsson, Luc Cuyvers, and et al. 2020. Research Is Needed to Inform Environmental Management of Hydrothermally Inactive and Extinct Polymetallic Sulfide (PMS) Deposits. *Marine Policy* 121: 104183. [CrossRef]

Van Dover, Cindy L. 2019. Inactive Sulfide Ecosystems in the Deep Sea: A Review. *Frontiers in Marine Science* 6: 461. [CrossRef]

Van Nijen, Kris, Steven Van Passel, Chris G. Brown, Michael W. Lodge, Kathleen Segerson, and Dale Squires. 2019. The Development of a Payment Regime for Deep Sea Mining Activities in the Area through Stakeholder Participation. *International Journal of Marine and Coastal Law* 34: 571–601. [CrossRef]

Vanreusel, Ann, Ana Hilario, Pedro A. Ribeiro, Lenaick Menot, and Pedro Martínez Arbizu. 2016. Threatened by Mining, Polymetallic Nodules Are Required to Preserve Abyssal Epifauna. *Scientific Reports* 6: 26808. [CrossRef]

Volz, Jessica B., Laura Haffert, Matthias Haeckel, Andrea Koschinsky, and Sabine Kasten. 2020. Impact of Small-Scale Disturbances on Geochemical Conditions, Biogeochemical Processes and Element Fluxes in Surface Sediments of the Eastern Clarion-Clipperton Zone, Pacific Ocean. *Biogeosciences* 17: 1113–31. [CrossRef]

Weaver, Philip P. E., and David Billet. 2019. Environmental Impacts of Nodule, Crust and Sulphide Mining: An Overview. In *Environmental Issues of Deep-Sea Mining: Impacts, Consequences and Policy Perspectives*. Edited by Rahul Sharma. Cham: Springer International Publishing, pp. 27–62.

Weaver, Philip P. E., David Billet, and Cindy L. Van Dover. 2018. Environmental Risks of Deep-Sea Mining. In *Handbook on Marine Environment Protection*. Edited by Markus Salomon and Till Markus. Cham: Springer International Publishing, pp. 215–45.

Wessel, Paul, David T. Sandwell, and Seung-Sep Kim. 2010. The Global Seamount Census. *Oceanography* 23: 24–33. [CrossRef]

MDPI

St. Alban-Anlage 66

4052 Basel

Switzerland

Tel. +41 61 683 77 34

Fax +41 61 302 89 18

www.mdpi.com

MDPI Books Editorial Office

E-mail: books@mdpi.com

www.mdpi.com/books

www.ingramcontent.com/pod-product-compliance
Lightning Source LLC
LaVergne TN
LVHW071927160726
843515LV00010B/2546